Application and Optimization of CCUS Technology in Shale Gas Production and Storage

Application and Optimization of CCUS Technology in Shale Gas Production and Storage

Guest Editors

Jun Liu

Gan Feng

Peng Zhao

MDPI

Basel • Beijing • Wuhan • Barcelona • Belgrade • Novi Sad • Cluj • Manchester

Guest Editors

Jun Liu
Institute of New Energy and Low-Carbon Technology
Sichuan University
Chengdu
China

Gan Feng
College of Water Resource & Hydropower
Sichuan University
Chengdu
China

Peng Zhao
College of Energy
Chengdu University of Technology
Chengdu
China

Editorial Office
MDPI AG
Grosspeteranlage 5
4052 Basel, Switzerland

This is a reprint of the Special Issue, published open access by the journal *Energies* (ISSN 1996-1073), freely accessible at: https://www.mdpi.com/journal/energies/special_issues/64H4B71801.

For citation purposes, cite each article independently as indicated on the article page online and as indicated below:

Lastname, A.A.; Lastname, B.B. Article Title. *Journal Name* **Year**, *Volume Number*, Page Range.

ISBN 978-3-7258-5129-4 (Hbk)
ISBN 978-3-7258-5130-0 (PDF)
https://doi.org/10.3390/books978-3-7258-5130-0

Contents

energies

Editorial

Application and Optimization of CCUS Technology in Shale Gas Production and Storage

Jun Liu [1], Gan Feng [2,*] and Peng Zhao [3]

1 Institute of New Energy and Low-Carbon Technology, Sichuan University, Chengdu 610065, China; j.liu@scu.edu.cn
2 State Key Laboratory of Hydraulics and Mountain River Engineering, College of Water Resource & Hydropower, Sichuan University, Chengdu 610065, China
3 State Key Laboratory of Geohazard Prevention and Geoenvironment Protection, Chengdu University of Technology, Chengdu 610059, China; scu_zhaopeng@163.com
* Correspondence: fenggan@whu.edu.cn

Global climate change is a crucial issue confronting the international community. The greenhouse effect caused by CO_2 emissions is widely recognized as one of the key factors impacting climate change [1–4]. Reducing CO_2 emissions and fully utilizing CO_2 are important ways to protect the environment and achieve resource development. Carbon dioxide capture, utilization, and storage technology (CCUS) is an emerging technology with the potential to result in large-scale CO_2 reduction; therefore, it has received widespread attention both domestically and internationally in the past decade. For example, the United States Department of Energy released the Carbon Dioxide Capture and Storage (CCS) Technology Research and Development Demonstration Roadmap in 2010, which laid out a development path for CCS and focused on efficient and economical solutions to quickly realize commercial applications. CCS technology needs to overcome various challenges at the economic, social, and technological levels, including the effective monitoring and verifying of CO_2 storage processes and the viability of the permanent underground storage of CO_2. The development of the Canadian CCS Technology Development Roadmap (CCSTRM) underwent four stages: situation analysis, confirmation of technology paths and R&D strategies, confirmation of priority directions for R&D and demonstration, and final writing. Japan's CCS research is basically synchronized with European and North American countries. In Japan, CCS technology was implemented in 2020, and encompasses various applications such as enhancing oil recovery using CO_2, sealing CO_2 up to saltwater layers, injecting CO_2 into depleted natural gas fields and coal seams, and developing CO_2 transportation technology. The goal in China is to achieve peak CO_2 emissions by 2030 and carbon neutrality by 2060 through stronger policies and measures. This further emphasizes the concept of "turning waste into wealth", and emphasizes opportunities to improve the recovery rates of oil, natural gas, coalbed methane, and water using carbon dioxide (CO_2-EOR, CO_2-EGR, CO_2-ECBM, and CO_2-EWR, respectively) [5].

Natural gas resources in unconventional sedimentary reservoirs, such as shale and coal seams rich in kerogen, are abundant and environmentally friendly, making them one of the most important energy resources available [6–11]. CO_2-enhanced shale gas production and underground storage integration technology are important ways to apply CCUS. Because of its characteristics of energy conservation, emission reduction, and efficient exploitation of shale gas, CCUS is emerging as an important future means for efficient shale gas exploitation and development [3,12,13]. At the same time, the stability of shale reservoirs is also a focus of research, directly determining the safety and efficiency of mining. The failure criterion that fully considers shale anisotropy and hydration is an important factor in evaluating the stability of shale formations [14]. The continuous exploitation and use of shallow shale gas is causing shale gas exploration and development to gradually move to deeper locations. These deeper environments have complex and changeable geological conditions, high temperatures, high ground stress, fluid seepage pressure, and other complex conditions [15–21]. This makes innovations in applying CCUS technology during deep shale gas exploitation urgent. CCUS application in deep shale gas exploitation is now an important research direction.

Scholars have studied shale pore structure, shale gas recovery, shale reservoir stability, and shale gas movement and transmission in the application of CCUS in shale gas exploitation. For example, Wang et al. [22] analyzed the displacement mechanism of CH_4 in nanoporous shale, its positive and negative effects on solid storage and organic matter extraction, the migration and expansion processes of elements, and the macro- and micro-control mechanisms for improving reservoir permeability. Feng et al. [23,24] studied the fracture trajectory of shale in the Longmaxi Formation after failure, and explored the mechanical properties of the shale under supercritical CO_2 adsorption conditions. Asif et al. [25] measured the isotherms of two gases (CH_4 and CO_2) at pressures up to 8.4 MPa, and constructed an isotherm of CO_2 to estimate the enhanced recovery rate of coalbed methane. Kong et al. [26] studied auxiliary additives to improve shale oil and gas production. Chen et al. [27] studied the creep characteristics of shale, considering the influence of bedding direction and water content.

The application of CCUS in shale gas production and storage involves many aspects, including in-depth research on its mechanisms and key technologies. This includes the competitive adsorption mechanism of CO_2/CH_4 in shale, SC-CO_2 displacement and the displacement of shale gas, the SC-CO_2 storage mechanism, SC-CO_2 jet technology, SC-CO_2 fracturing technology, the CO_2 displacement of shale gas, and SC-CO_2 storage technology. These issues need to be analyzed and solved from the perspectives of technology, security, investment, environment, and ecology. Therefore, we have organized this Special Issue of *Energy*, titled "Application and Optimization of CCUS Technology in Shale Gas Production and Storage". This Special Issue welcomes contributions of all achievements regarding CCUS technology related to shale gas, including all outlines from laboratory experiments, numerical simulations, engineering evaluations, and economic judgement.

Author Contributions: J.L, G.F. and P.Z. contributed equally to the design, implementation, conceptualization, and the delivery of the Special Issue; writing—original draft preparation, G.F.; writing—review and editing, J.L. and G.F. All authors have read and agreed to the published version of the manuscript.

Funding: This study was supported by the National Natural Science Foundation of China (Grant No. 52104143), the Natural Science Foundation of Sichuan Province, China (Grant No. 2022NSFSC0193), and the Open Fund of State Key Laboratory of Water Resource Protection and Utilization in Coal Mining (Grant No. GJNY-21-41-01), which are greatly appreciated.

Conflicts of Interest: The authors declare no conflict of interest.

References

1. Gilliland, E.S.; Ripepi, N.; Conrad, M.; Miller, M.J.; Karmis, M. Selection of monitoring techniques for a carbon storage and enhanced coalbed methane recovery pilot test in the Central Appalachian Basin. *Int. J. Coal Geol.* **2013**, *118*, 105–112. [CrossRef]
2. Kelemen, P.; Benson, S.M.; Pilorgé, H.; Psarras, P.; Wilcox, J. An Overview of the Status and challenges of CO_2 Storage in Minerals and Geological Formations. *Front. Clim.* **2019**, *1*, 9. [CrossRef]
3. Liu, J.; Yao, Y.; Liu, D.; Elsworth, D. Experimental evaluation of CO_2 enhanced recovery of adsorbed-gas from shale. *Int. J. Coal Geol.* **2017**, *179*, 211–218. [CrossRef]
4. Shi, F.; Deng, B.; Yin, G.; Zhang, D.; Li, M.; Liu, P.; Liu, C. Kinetic behavior of heterogeneous sorption deformation on coal: Effect of maceral/micro-lithotype distribution. *Int. J. Coal Geol.* **2019**, *216*, 103324. [CrossRef]
5. Jiang, K.; Ashworth, P. The development of Carbon Capture Utilization and Storage (CCUS) research in China: A bibliometric perspective. *Renew. Sust. Energy Rev.* **2021**, *138*, 110521. [CrossRef]
6. Kang, J.; Wan, R.; Zhou, F.; Liu, Y.; Li, Z.; Yin, Y. Effects of supercritical CO_2 extraction on adsorption characteristics of methane on different types of coals. *Chem. Eng. J.* **2019**, *388*, 123449. [CrossRef]
7. Huang, L.; Ning, Z.; Wang, Q.; Qi, R.; Cheng, Z.; Wu, X.; Zhang, W.; Qin, H. Molecular Insights into Kerogen Deformation Induced by CO_2/CH_4 Sorption: Effect of Maturity and Moisture. *Energy Fuel* **2019**, *33*, 4792–4805. [CrossRef]

8. Li, J.; Wang, Y.; Chen, Z.; Rahman, S.S. Simulation of Adsorption–Desorption Behavior in Coal Seam Gas Reservoirs at the Molecular Level: A Comprehensive Review. *Energy Fuel* **2020**, *34*, 2619–2642. [CrossRef]
9. Meng, M.; Qiu, Z.; Zhong, R.; Liu, Z.; Liu, Y.; Chen, P. Adsorption characteristics of supercritical CO_2/CH_4 on different types of coal and a machine learning approach. *Chem. Eng. J.* **2019**, *368*, 847–864. [CrossRef]
10. Mohanty, M.M.; Pal, B.K. Sorption behavior of coal for implication in coal bed methane an overview. *Int. J. Min. Sci. Technol.* **2017**, *27*, 307–314. [CrossRef]
11. Li, J.; Sun, C.J.E. Molecular insights on competitive adsorption and enhanced displacement effects of CO_2/CH_4 in coal for low-carbon energy technologies. *Energy* **2022**, *261*, 125176. [CrossRef]
12. Cheng, L.; Li, D.; Wang, W.; Liu, J. Heterogeneous Transport of Free CH_4 and Free CO_2 in Dual-Porosity Media Controlled by Anisotropic In Situ Stress during Shale Gas Production by CO_2 Flooding: Implications for CO_2 Geological Storage and Utilization. *ACS Omega* **2021**, *6*, 26756–26765. [CrossRef] [PubMed]
13. Liu, J.; Xie, L.; Elsworth, D.; Gan, Q. CO_2/CH_4 Competitive Adsorption in Shale: Implications for Enhancement in Gas Production and Reduction in Carbon Emissions. *Environ. Sci. Technol.* **2019**, *53*, 9328–9336. [CrossRef] [PubMed]
14. Zhang, Q.; Yao, B.; Fan, X.; Li, Y.; Fantuzzi, N.; Ma, T.; Chen, Y.; Zeng, F.; Li, X.; Wang, L. A failure criterion for shale considering the anisotropy and hydration based on the shear slide failure model. *Int. J. Min. Sci. Technol.* **2023**, *33*, 447–462. [CrossRef]
15. Feng, G.; Wang, X.; Kang, Y.; Zhang, Z. Effect of thermal cycling-dependent cracks on physical and mechanical properties of granite for enhanced geothermal system. *Int. J. Rock Mech. Min. Sci.* **2020**, *134*, 104476. [CrossRef]
16. Zhang, A.; Xie, H.; Zhang, R.; Ren, L.; Zhou, J.; Gao, M.; Tan, Q. Dynamic failure behavior of Jinping marble under various preloading conditions corresponding to different depths. *Int. J. Rock. Mech. Min. Sci.* **2021**, *148*, 104959. [CrossRef]
17. Hu, J.; Xie, H.; Gao, M.; Li, C.; Sun, Q. Damage mechanism and heat transfer characteristics of limestone after thermal shock cycle treatments based on geothermal development. *Int. J. Rock. Mech. Min. Sci.* **2022**, *160*, 105269. [CrossRef]
18. Feng, G.; Wang, X.; Wang, M.; Kang, Y. Experimental investigation of thermal cycling on fracture characteristics of granite in a geothermal-energy reservoir. *Eng. Fract. Mech.* **2020**, *235*, 107180. [CrossRef]
19. Liu, C.; Feng, G.; Xie, H.; Wang, J.; Duan, Z.; Tao, Y.; Lu, G.; Xu, H.; Hu, Y.; Li, C.; et al. Study on the accuracy of fracture criteria in predicting fracture characteristics of granite with different occurrence depths. *Energies* **2022**, *15*, 9248. [CrossRef]
20. Feng, G.; Zhu, C.; Wang, X.; Tang, S. Thermal effects on prediction accuracy of dense granite mechanical behaviors using modified maximum tangential stress criterion. *J. Rock Mech. Geotech. Eng.* **2023**, *15*, 1734–1748. [CrossRef]
21. Jin, P.; Hu, Y.; Shao, J.; Zhao, G.; Zhu, X.; Li, C. Influence of different thermal cycling treatments on the physical, mechanical and transport properties of granite. *Geothermics* **2019**, *78*, 118–128. [CrossRef]
22. Wang, M.; Huang, K.; Xie, W.; Dai, X. Current research into the use of supercritical CO_2 technology in shale gas exploitation. *Int. J. Min. Sci. Technol.* **2019**, *29*, 739–744. [CrossRef]
23. Feng, G.; Kang, Y.; Wang, X.; Hu, Y.; Li, X. Investigation on The Failure Characteristics and Fracture Classification of Shale Under Brazilian Test Conditions. *Rock Mech. Rock Eng.* **2020**, *53*, 3325–3340. [CrossRef]
24. Feng, G.; Kang, Y.; Sun, Z.-D.; Wang, X.-C.; Hu, Y.-Q. Effects of supercritical CO_2 adsorption on the mechanical characteristics and failure mechanisms of shale. *Energy* **2019**, *173*, 870–882. [CrossRef]
25. Asif, M.; Panigrahi, D.; Naveen, P.; Ojha, K. Construction of high-pressure adsorption isotherm: A tool for predicting coalbed methane recovery from Jharia coalfield, India. *Int. J. Min. Sci. Technol.* **2019**, *29*, 765–769. [CrossRef]
26. Kong, S.; Feng, G.; Liu, Y.; Li, K. Potential of dimethyl ether as an additive in CO_2 for shale oil recovery. *Fuel* **2021**, *296*, 120643. [CrossRef]
27. Chen, Y.; Hao, X.; Teng, J.; Wen, Z. Creep behavior of water-containing bedded shale. *J. Cent. South. Univ.* **2023**, *30*, 975–991. [CrossRef]

energies

MDPI

Article

Multi-Fracture Propagation Considering Perforation Erosion with Respect to Multi-Stage Fracturing in Shale Reservoirs

Lin Tan [1,*], Lingzhi Xie [2], Bo He [2] and Yao Zhang [1]

[1] College of Architecture and Environment, Sichuan University, Chengdu 610065, China; zhangyao@scu.edu.cn
[2] Institute of New Energy and Low-Carbon Technology, Sichuan University, Chengdu 610065, China; xielingzhi@scu.edu.cn (L.X.); hebo@scu.edu.cn (B.H.)
* Correspondence: tanscu@163.com

Abstract: Shale gas is considered a crucial global energy source. Hydraulic fracturing with multiple fractures in horizontal wells has been a crucial method for stimulating shale gas. During multi-stage fracturing, the fracture propagation is non-uniform, and fractures cannot be induced in some clusters due to the influence of stress shadow. To improve the multi-fracture propagation performance, technologies such as limited-entry fracturing are employed. However, perforation erosion limits the effect of the application of these technologies. In this paper, a two-dimensional numerical model that considers perforation erosion is established based on the finite element method. Then, the multi-fracture propagation, taking into account the impact of perforation erosion, is studied under different parameters. The results suggest that perforation erosion leads to a reduction in the perforation friction and exacerbates the uneven propagation of the fractures. The effects of erosion on multi-fracture propagation are heightened with a small perforation diameter and perforation number. However, reducing the perforation number and perforation diameter remains an effective method for promoting uniform fracture propagation. As the cluster spacing is increased, the effects of erosion on multi-fracture propagation are aggravated because of the weakened stress shadow effect. Furthermore, for a given volume of fracturing fluid, although a higher injection rate is associated with a shorter injection time, the effects of erosion on the multi-fracture propagation are more severe at a high injection rate.

Keywords: perforation erosion; shale gas; multi-fracture propagation; finite element method

1. Introduction

As conventional oil and gas production declines and energy demand increases, shale gas is regarded as an essential global energy source [1]. According to the evaluation results of the Energy Information Administration (EIA), the worldwide recoverable shale gas resources constitute approximately 206.68×10^{12} m^3, which have great potential for exploitation [2]. And as a low-carbon energy and clean source, increasing the use of shale gas contributes to reducing greenhouse gas emissions and air pollution [3,4]. However, due to the low permeability and porosity of shale reservoirs, the recovery of shale gas remains challenging [5,6]. Recently, multi-stage fracturing in horizontal wells has emerged as the most effective method for stimulating shale reservoirs [7–9]. During multi-stage fracturing, multiple fractures are created simultaneously to increase the volume stimulated in the reservoir [10–12]. Nevertheless, Miller et al. [13] analyzed the production logging data on a good deal of shale horizontal wells and concluded that approximately 33% of the perforation clusters were the main source of production in certain basins, while nearly another 33% of the perforation clusters produced nothing. According to optical fiber diagnosis, Ugueto C. et al. [14] found that the efficiency of perforation clusters was only 50~70%. The studies mentioned above suggest that fractures are not propagated uniformly in multi-stage fracturing, which is not conducive to reservoir stimulation [15,16].

Therefore, deeply comprehending the multi-fracture propagation mechanism is important for enhancing the recovery of shale gas.

During multi-stage fracturing, the morphology of the fractures is dominated by the stress shadow effect and the flow distribution between fractures. The flow distribution between fractures is determined by the resistance to fracture propagation, the perforation friction, and the pressure loss in the wellbore [17]. Nevertheless, the stress shadow between fractures leads to an increase in the propagation resistance of the interior fractures [18]. As fluid invariably flows through the path of least resistance, more fluid is distributed to the exterior fractures. Therefore, the effect of stress shadow should be weakened, or the increased resistance caused by stress shadow should be balanced, to promote a uniform flow distribution between fractures. Quite a few technologies have been proposed to promote uniform fracture propagation, such as limited-entry fracturing, temporary plugging fracturing, and fracture optimization [19–21]. Wu et al. [22] investigated multi-fracture propagation and found that reducing the perforation number and adjusting non-uniform cluster spacing could effectively improve the unevenness of the fracture propagation using the displacement discontinuity method (DDM). Peirce and Bunger [23] proposed a parallel planar 3D model and demonstrated that the stress shadow between fractures could result in uneven fracture propagation. Chen et al. [24] analyzed the flow distribution between fractures and performed sensitivity analysis using a semi-analytical model. Zhang et al. [25] studied extreme limited-entry fracturing (XLE) and optimized the controlling parameters influencing the cluster effectiveness. Li et al. [26] considered the impact of the perforation friction by establishing a fluid pipe element and analyzed the three-dimensional multi-fracture propagation. Their research noted that the number of fractures has a significant effect on multi-fracture propagation.

As stated above, reasonable optimization of the perforation parameters can effectively unify the fracture propagation. However, even when limited-entry fracturing has been applied in practical applications, the fractures can only temporarily maintain uniform fracture propagation, and severe non-uniform fracture propagation ultimately occurs [27,28]. Perforation erosion is one of the significant factors leading to this phenomenon [29,30]. Roberts et al. [31] used downhole photographic equipment to observe and record the size and shape of perforations after perforation erosion and found that the erosion of some perforations occurred faster than that of others. Cramer et al. [32] found that the perforation erosion in fractures in the heel of the wells was more severe compared with the fractures in the toe of the wells based on video imaging. The studies above suggest that perforation erosion is common during the process of hydraulic fracturing. Crump and Conway [30] proposed that abrasive damage to perforations consists of two mechanisms based on their experiments: one is an increase in the perforation diameter and another is the smoothing of the perforations. The smoothing of the perforations and an increase in the perforation diameter results in a reduction in the perforation friction, which results in a redistribution of fluid between the fractures and exerts a significant impact on the fracture propagation. Hence, understanding multi-fracture propagation under the effect of perforation erosion is of great importance to the effective design of multi-stage fracturing. However, perforation erosion is seldom considered in studies on multi-fracture propagation, yet perforation friction is generally considered a constant.

In this study, a numerical model considering perforation erosion is established. Multi-fracture propagation, taking into account the effect of perforation erosion under different conditions, is discussed in detail, including the perforation diameter, perforation number, cluster spacing, and injection rate. The key factors influencing the performance of multi-fracture propagation are summarized. This study can help us to understand the mechanisms of perforation erosion affecting multi-fracture propagation and provide recommendations for multi-stage fracturing design.

2. Numerical Model

In this section, a smeared crack model based on the finite element method (FEM) is developed. Currently, the numerical methods for simulating hydraulic fracturing can be divided into continuum approaches and discontinuum approaches. The cohesive zone model (CZM) is an intuitive discontinuum method, which simulates the failure and damage of material by specifying the traction–separation law [33]. However, the paths of the HFs and NFs must be defined in advance. The discrete element method (DEM) can directly simulate the interaction between fractures, but this method cannot be applied at the engineering scale due to its high computational cost. In addition to discontinuum methods, continuum approaches are another way to simulate hydraulic fracturing [34]. To solve the problem of the remeshing, the extended finite element method (XFEM) simulates fracture propagation by introducing an appropriate enrichment function [35]. Nevertheless, it is difficult to solve the multi-fracture problem, and the calculation efficiency is not high [36]. The smeared crack model based on the FEM is easy to use for simulating multi-fracture propagation, and the fractures are simplified as damage elements [37]. The method has an adequate efficiency in simulating the multi-fracture propagation in reservoirs. The model is described as follows:

2.1. Rock Deformation and Fluid Flow

The rock deformation equation can be expressed as [38]:

$$\overline{\nabla}\sigma + \gamma_b g_v - \rho\frac{d^2u}{dt^2} = f \tag{1}$$

where g_v is the gravity vector; $\overline{\nabla}$ is the matrix differentiation operator; γ_b is the bulk weight; σ is the total Cauchy stress; f is the force; ρ is the density; and u is the displacement.

The fluid inside the fractures satisfies the mass conservation law [38]:

$$\nabla^T q + \frac{1}{Q^*}\frac{dp_w}{dt} + \alpha\frac{d\varepsilon_v}{dt} = s \tag{2}$$

where p_w is the pore pressure; ∇ is the gradient operator; Q^* is the compressibility parameter; q is the flow flux vector; s is the flow source; and ε_v is the rock volumetric strain.

Darcy's law is applied to describe the relationship between the flow flux and pore pressure [39]:

$$q = \frac{K}{\gamma_f}\left(-\nabla p + \gamma_f g_v\right) \tag{3}$$

where γ_f is the bulk weight and K is the second-order permeability tensor.

2.2. Fracture Propagation

There are two propagation modes during the process of fracture propagation: one is toughness-dominated, in which rock fracture is the primary mechanism of energy dissipation, and another is viscosity-dominated, in which fluid viscosity dissipation is the primary mechanism of energy dissipation [39,40]. In large-scale hydraulic fracturing, the fracture propagation is considered to be viscosity-dominated. In this paper, since we are focusing on large-scale fracture propagation, the propagation of the fractures is determined by the maximum tensile stress criterion. Fractures are generated when the minimum effective principal stress equals the tensile strength:

$$\sigma_n = T \tag{4}$$

where T is the tensile strength and σ_n is the effective normal stress.

In this model, the fracture is represented by damage zones with enhanced permeability. The strain of a damaged element is composed of intact rock strain and fracture strain based on the continuum equivalent method. And the fracture strain and intact rock strain are approximated as the plastic strain and elastic strain of the elements, respectively, as shown in Figure 1. Hence, the plastic strain of the elements is used to calculate the equivalent fracture width:

$$w_f = L_e \cdot \varepsilon_p (\delta_{ij} - n_i n_j) \tag{5}$$

where ε_p is the plastic strain tensor and L_e is the element size.

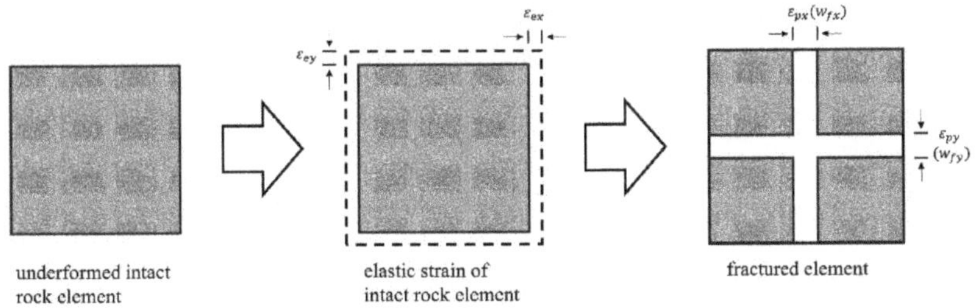

Figure 1. Schematic illustration of deformation of a rock element.

The permeability of a damaged element is composed of matrix permeability and fracture permeability. In general, the deformation of intact rock has less effect on the matrix permeability [41]. Hence, the permeability of the matrix is considered to be constant, and the increase in the damaged element's permeability is completely due to the fracture propagation. The fracture permeability can be calculated according to the cubic law:

$$k_f = \frac{\rho g}{L_e} \frac{w_f^3}{12u} \tag{6}$$

where u is the fluid viscosity and ρ is the fluid density.

The permeability of a damaged element can be written as:

$$k = k_m + k_f \tag{7}$$

where k_m is the matrix permeability.

2.3. Fluid Flow within the Wellbore

During multi-stage fracturing, stress shadow can result in differences in the injection rates between fractures. Hence, the dynamic flow distribution between the fractures should be considered. Because the horizontal wellbores and perforations are not smooth, frictional resistance is generated as fluid flows through the horizontal wellbores and perforations. Generally, the friction resistance generated by the wellbores is small and can be ignored [42]. According to the pressure balance conditions, the relationship between the pressure at the horizontal well heel, the pressure at the fracture entrance, and the perforation friction pressure drop can be expressed as:

$$p_{b,1} = \cdots = p_{b,i} = \cdots = p_{b,m}$$

$$\begin{cases} p_{b,1} = p_{p,1} + p_{f,1} \\ \vdots \\ p_{b,i} = p_{p,i} + p_{f,i} \\ \vdots \\ p_{b,m} = p_{p,m} + p_{f,m} \end{cases} \tag{8}$$

where $p_{b,i}$, $p_{f,i}$, and $p_{p,i}$ are the pressure at the horizontal well heel, the pressure at the entrance of the ith fracture, and the perforation friction pressure drop, respectively.

The perforation friction pressure drop can be expressed as [43]:

$$p_p = \alpha_f q_p^2 \tag{9}$$

$$\alpha_f = \frac{8\rho}{\pi^2 C_d^2 n_p^2 d_p^4} \tag{10}$$

where α_f is the perforation friction coefficient; n_p is the perforation number; ρ is the slurry density; d_p is the perforation diameter; q_p is the flow rate; and C_d is the discharge coefficient.

Based on volume conservation, the total injection rate equals the injection rate of each fracture:

$$Q = \sum_{i=1}^{m} q_i \tag{11}$$

2.4. Perforation Erosion

Abrasive damage to the perforations occurs when the proppant is pumped at a high injection rate. Crump and Conway [30] found that perforation erosion is mainly caused by two mechanisms: one is the rounding and smoothing of perforations, and another is an increase in the perforation diameter, as shown in Figure 2. Based on the experimental data, Long et al. [44] established a semi-empirical perforation erosion model:

$$\frac{\delta d_p}{\delta t} = \overline{\alpha} C_e v^2 \tag{12}$$

$$\frac{\delta C_d}{\delta t} = \overline{\beta} C_e v^2 \left(1 - \frac{C_d}{C_d^{max}}\right) \tag{13}$$

where d_p is the perforation diameter; v is the mean velocity at the perforations; C_e is the proppant concentration; C_d^{max} is the maximum discharge coefficient (the range of C_d is 0.6–0.9); and $\overline{\alpha}$ and $\overline{\beta}$ are empirical coefficients based on laboratory experiments.

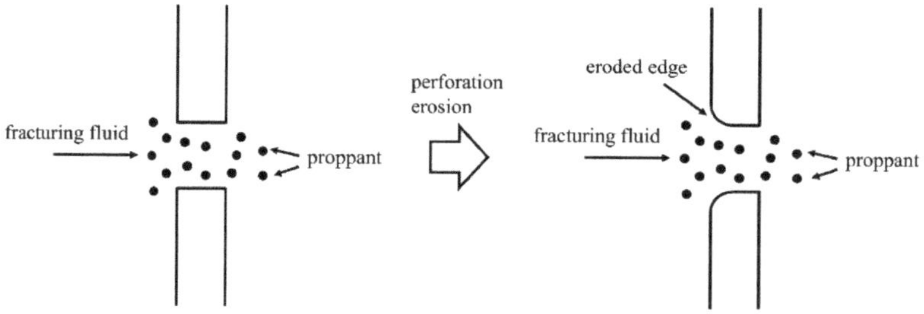

Figure 2. Schematic of perforation abrasion.

The mean velocity at the perforation can be expressed as:

$$v = \frac{4q_p}{n_p \pi d_p} \tag{14}$$

3. Model Verification and Model Establishment

3.1. Comparison with the Classical KGD Model

In this section, the classical KGD model is applied to verify the fracture propagation in the numerical model. The fracture width and length in the KGD model can be expressed as [45]:

$$L = 0.539 \left(\frac{Q^3 E}{4(1-v^2)\mu H^3}\right)^{\frac{1}{6}} t^{\frac{2}{3}} \tag{15}$$

$$w = 2.36\left(\frac{4(1-v^2)\mu Q^3}{EH^3}\right)^{\frac{1}{6}} t^{\frac{1}{3}} \quad (16)$$

where Q is the injection rate; v is Poisson's ratio; E is Young's modulus; t is the injection time; H is the fracture height; and μ is the fluid viscosity. Table 1 shows the parameters utilized in the validation. And Figure 3 illustrates a comparison between the KGD model and the numerical model. The numerical results and the theoretical results were in great agreement.

Table 1. Basic parameters for the model validation.

Parameter	Value
Young's modulus	20 GPa
Poisson's ratio	0.2
Tensile strength	6 MPa
Injection rate	0.001 m^3/s
Viscosity	70 cP
Pay thickness	1 m
Minimum horizontal stress	7 MPa

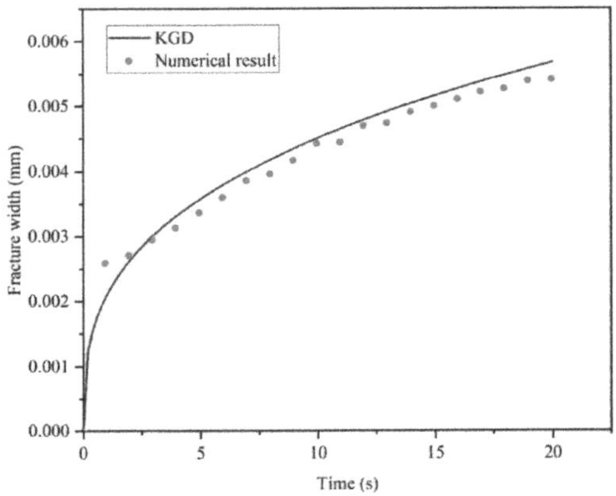

(a)

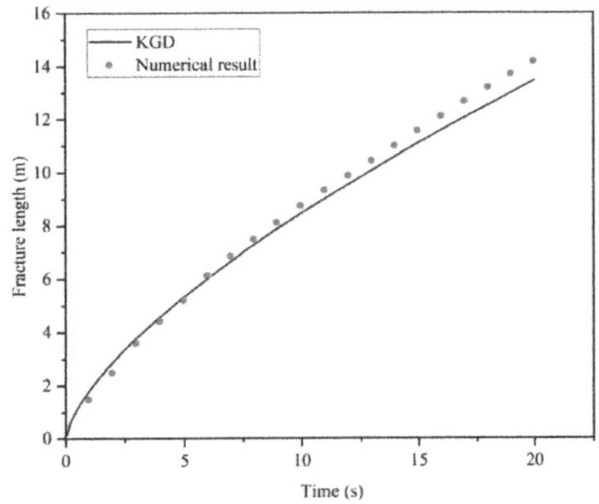

(b)

Figure 3. Comparison of numerical results and KGD model: (**a**) comparison of fracture width; (**b**) comparison of fracture length.

3.2. Comparison with a Lab-Scale Experiment

In this section, the numerical model is verified by comparing it with the experimental results of Xing et al. [46]. Their experiment monitored the propagation of a hydraulic fracture, as shown in Figure 4a. The size of the specimen was 152.4 mm × 152.4 mm × 114.3 mm, and the specimen consisted of six transparent polyurethane blocks, with three layers included. The specimen was loaded under true tri-axial conditions, and a higher stress was applied to the barrier layers to restrict the propagation of the fracture in the height direction. The fluid was injected at a constant injection rate throughout the experiment, and the propagation of fractures was monitored using a video camera. The photometric method was used to obtain the fracture aperture, and thus the fracture opening along the fracture length was obtained. Table 2 lists the parameters used in the experiment. Figure 4b shows a comparison of the numerical model and the experimental results. The numerical model is in great agreement with the experimental results.

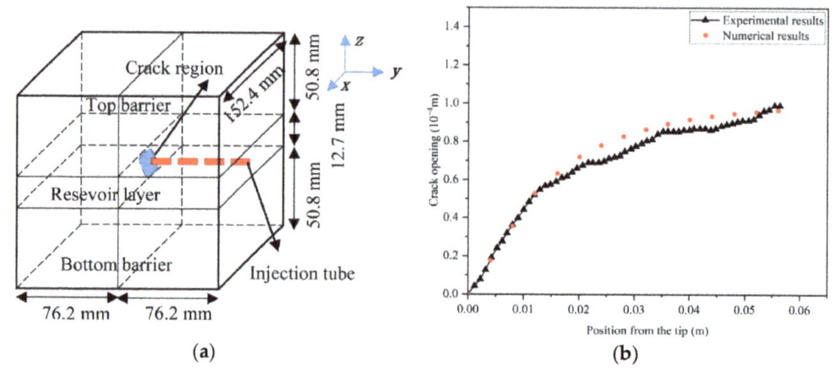

Figure 4. (**a**) Sketch of experiments; (**b**) comparison between numerical results and experimental results.

Table 2. Basic parameters for the model validation.

Parameter	Value
Young's modulus	0.1 GPa
Poisson's ratio	0.48
Injection rate	8 mL/min
Viscosity	300 cP
Minimum horizontal stress	0.31 MPa

3.3. Model Establishment

To investigate multi-fracture propagation, a 120 × 90 m two-dimensional model for multi-fracture propagation is established. Three clusters are considered in the model, and the horizontal wellbore is oriented in the direction of the minimum horizontal principal stress. The model contains a total of 10,800 elements. The displacement boundary conditions in the model are illustrated in Figure 5. The constraints for displacement in the x-direction are applied to the left and right boundaries, while the constraints for displacement in the y-direction are applied to the upper and lower boundaries.

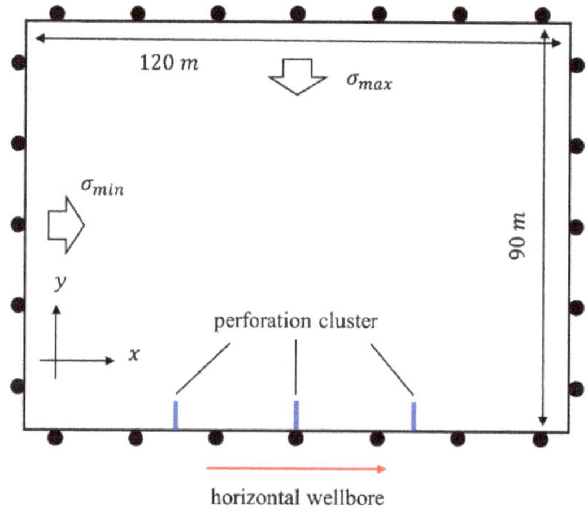

Figure 5. Schematic of simulation model.

4. Results and Discussion

In this section, multi-fracture propagation considering the perforation erosion is compared with that without considering the perforation erosion to illustrate the impact of perforation erosion with respect to competitive multi-fracture propagation under different parameters, including injection rate, perforation number, perforation diameter, and cluster spacing. Table 3 presents the basic parameters in the model.

Table 3. Basic parameters for the simulation.

Parameter	Value
Young's modulus	15 GPa
Poisson's ratio	0.2
Density	2400 kg/m^3
Tensile strength	6 MPa
Biot coefficient	0.7
Injection rate	0.1 m^3/s
Viscosity	1 cP
Maximum horizontal stress	71 MPa
Minimum horizontal stress	63 MPa
Perforation diameter	10 mm
Perforation number	12
Cluster spacing	25 m
Initial discharge coefficient	0.6
Maximum discharge coefficient	0.9
Sand concentration	720 kg/m^3

4.1. The Impact of Perforation Erosion on Multi-Fracture Propagation

In this section, the effect of perforation erosion in relation to competitive multi-fracture propagation is studied. The cluster spacing is set at 35 m, and the perforation diameter is set at 8 mm; the additional parameters are detailed in Table 2.

A comparison of the fracture propagation morphology between the cases with consideration of the perforation erosion and the cases without consideration of the perforation erosion is shown in Figure 6. The unevenness of the fracture propagation is intensified when considering perforation erosion. This is because perforation erosion gives rise to an increase in the perforation diameter (D) and discharge coefficient (C_d), as shown in Figure 7. According to Equation (10), the increase in the discharge coefficient and perforation diameter can lead to a reduction in the perforation friction, which aggravates an uneven flow distribution between fractures and uneven fracture propagation.

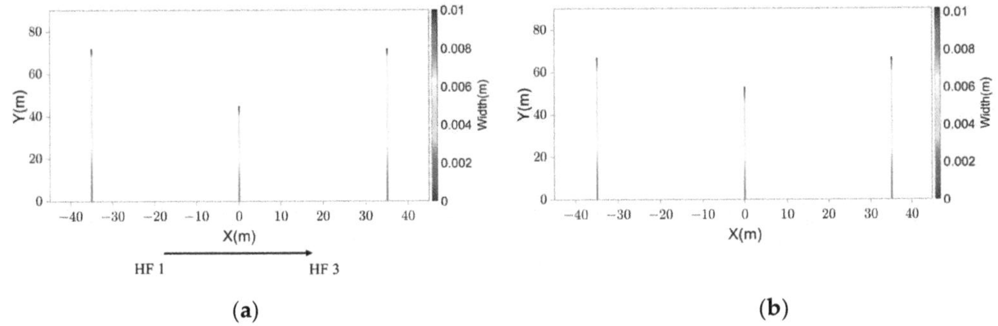

Figure 6. Fracture propagation under the two cases: (**a**) erosion; (**b**) no erosion.

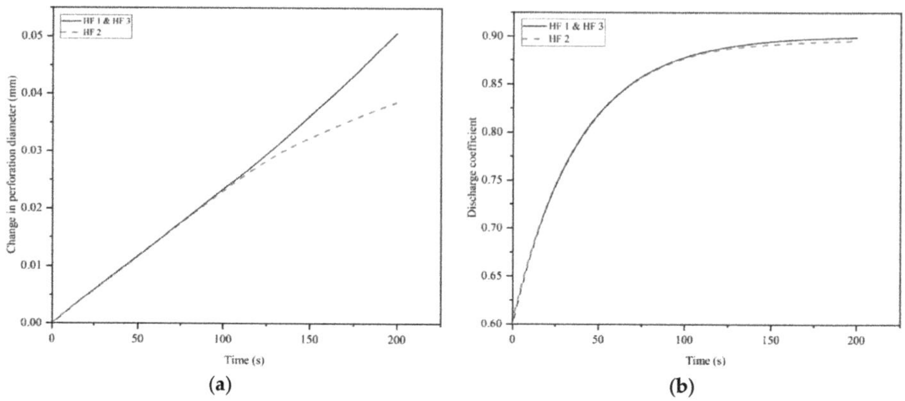

Figure 7. The change in the perforation diameter and discharge coefficient: (**a**) perforation diameter; (**b**) discharge coefficient.

The variation in the perforation friction coefficient in the process of perforation erosion is depicted in Figure 8. The perforation friction coefficient decreases rapidly at first and then decreases slowly. This is because the discharge coefficient increases rapidly from the initial value of 0.6 to 0.9 during the initial stage of perforation erosion, which causes the perforation friction coefficient to decrease rapidly in a short time. And then the discharge coefficient tends to remain stable, while the perforation diameter increases at a slow rate, which results in a slow decline in the perforation friction coefficient during the later stage. Hence, the rapid increase in the discharge coefficient is the main factor that aggravates non-uniform fracture propagation during the initial stage, while the increase in the perforation diameter becomes the main factor that aggravates non-uniform fracture propagation during the later stage.

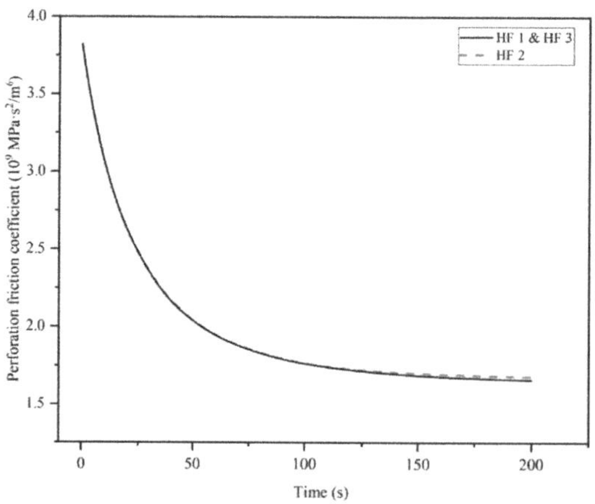

Figure 8. The change in perforation friction coefficient.

In summary, perforation erosion increases the perforation diameter and discharge coefficient, which reduces the frictional resistance generated by perforation and aggravates uneven fracture propagation. And the increases in the perforation diameter and discharge coefficient are the primary factors aggravating uneven fracture propagation during the initial and later stage, respectively.

4.2. The Effects of the Injection Rate

The injection rate is one of the crucial operation parameters that influences fracture propagation. When the proppant is pumped at a high speed, the effect of perforation erosion is unneglectable [47]. To study multi-fracture propagation under different injection rates, four cases are included (0.08 m^3/s, 0.1 m^3/s, 0.12 m^3/s, and 0.14 m^3/s). The injection volume remains consistent in all cases.

Figure 9 illustrates the fracture morphology at four different injection rates. As the injection rate increases, the uneven fracture propagation is improved. However, the length of the central fracture continues to increase, while the length of the side fractures changes slightly as the injection rate increases, as shown in Figure 10. This is because the fracturing fluid volume injected into the fractures remains constant, but the injection time is shorter at a higher injection rate. Figure 11 illustrates that the total injection volume distributed to the fractures on both sides decreases at an increasing injection rate, which tends to cause a reduction in the length of the side fractures. Nevertheless, a decrease in the injection time causes the injection rate into the fractures on both sides to increase. A higher injection rate contributes to an increase in the fracture length [48,49]. Hence, the length of the side fractures changed slightly.

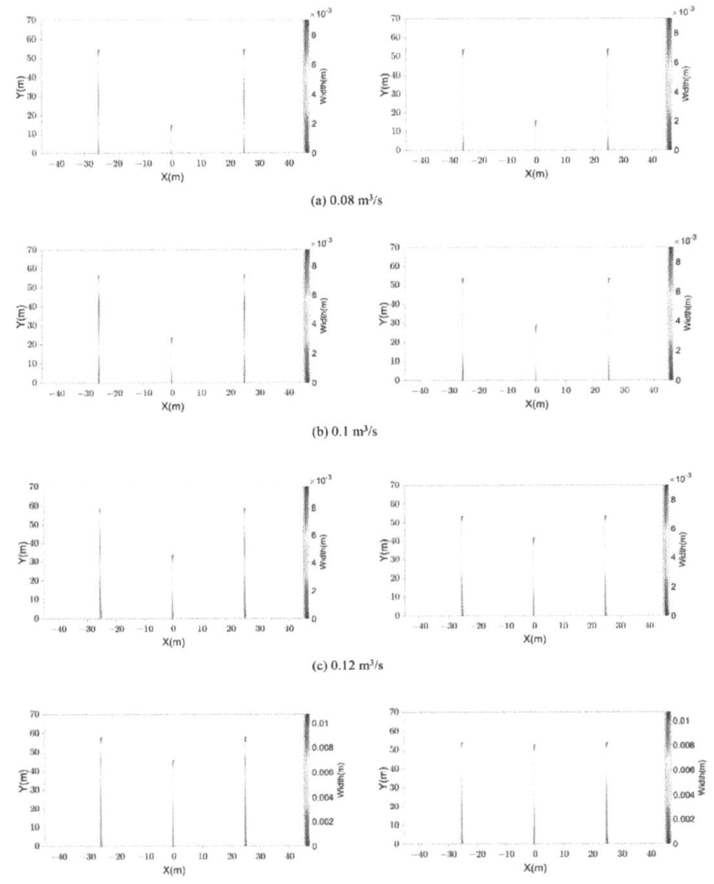

Figure 9. Fracture propagation under different injection rates (left: erosion; right: no erosion).

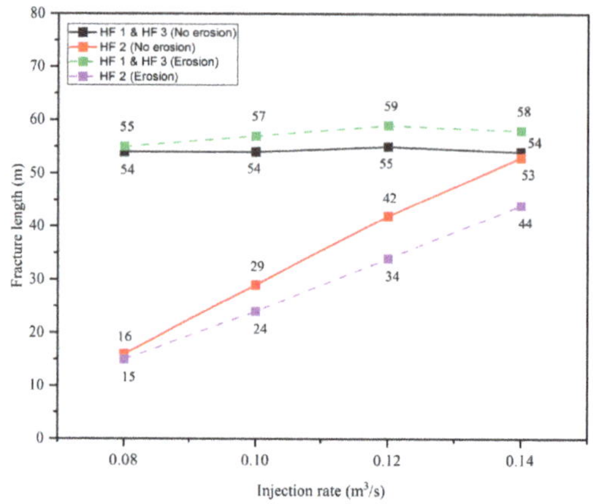

Figure 10. The length of each fracture under different injection rates.

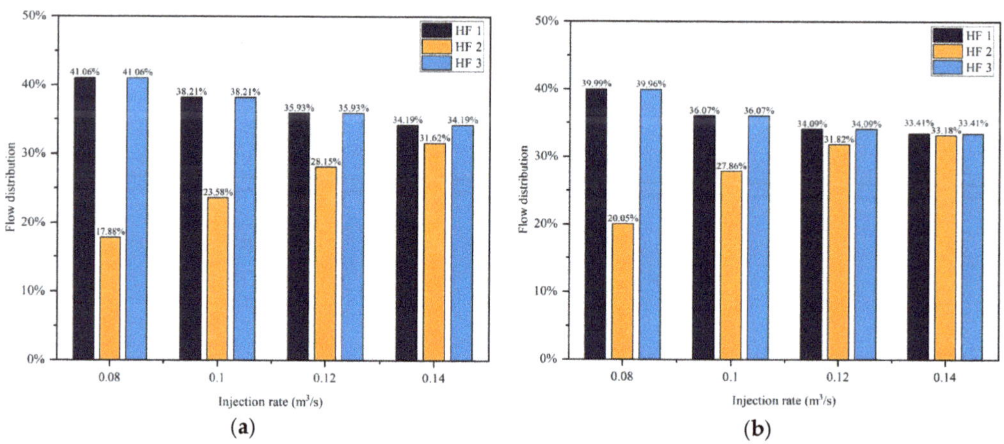

Figure 11. The flow distribution between fractures under different injection rates: (**a**) erosion; (**b**) no erosion.

Figure 12 shows the variation in the perforation diameter and discharge coefficient under different injection rates. The variation in the diameter and discharge coefficient of the perforations is greater as the injection rate increases within the same time. According to Equation (14), the mean velocity at the perforation increases as the injection rate increases, which results in a greater variation in the diameter and discharge coefficient of the perforations. Consequently, the perforation erosion is heightened at an increasing injection rate. Furthermore, the volume of fracturing fluid remains constant, although the injection time is longer at a lower injection rate, the perforation erosion becomes more severe at a higher injection rate, as illustrated in Figure 12. And due to this, Figure 10 shows that the difference in the central fracture length between the cases considering perforation erosion and the cases without considering perforation erosion is greater when the injection rate increases, which suggests that the effects of erosion on multi-fracture propagation are more severe at a higher injection rate.

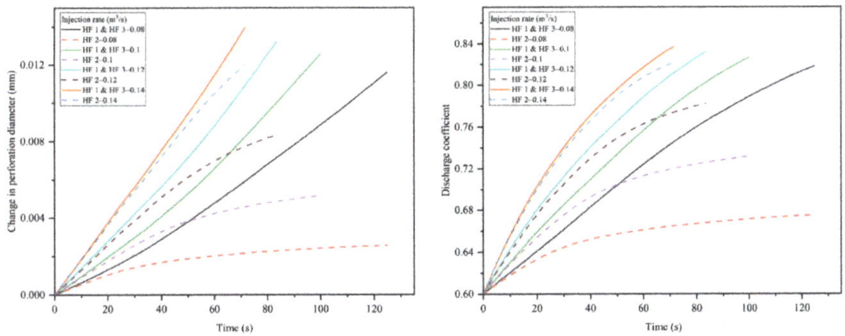

(a) (b)

Figure 12. The change in perforation diameter and discharge coefficient under different injection rates: (**a**) perforation diameter; (**b**) discharge coefficient.

In addition, the variation in the discharge coefficient and perforation diameter in the fractures on both sides is larger than that in the central fracture; the perforation erosion in the side fractures is thus more severe, as shown in Figure 12. This is because the amount of fluid distributed to each fracture is different. The injection rate into the fractures on both sides is larger and the injection rate into the central fracture is smaller because of the influence of stress shadow, as shown in Figure 11. A higher injection rate results in more significant perforation erosion in the side fractures. And the difference in the perforation erosion between the side fractures and the central fracture decreases as the injection rate increases. Figure 11 illustrates that the difference in the injection rate between the side fractures and the central fracture decreases as the total injection rate increases, which minimizes the difference in the perforation erosion between the side fractures and the central fracture.

Generally, for a given volume of fracturing fluid, the effects of erosion on multi-fracture propagation are more severe at a higher injection rate. And a decrease in the injection rate can mitigate the perforation erosion. However, increasing the injection rate can still contribute to uniform fracture propagation even though the effects of erosion on multi-fracture propagation are aggravated at an increasing injection rate. Therefore, a higher injection rate is recommended to promote uniform fracture propagation.

4.3. The Effects of the Perforation Diameter

The perforation diameter is a key factor influencing multi-fracture propagation. To investigate multi-fracture propagation under different perforation diameters, four cases are set in the model (8 mm, 9 mm, 10 mm, and 11 mm).

Figure 13 shows the fracture propagation at different perforation diameters. As the perforation diameter decreases, the length of the fractures on both sides continues to decrease, while the central fracture length continues to increase. And Figure 14 shows that the uneven flow distribution between the fractures is improved as the perforation diameter decreases. A reduction in the perforation diameter contributes to improving non-uniform fracture propagation. Nevertheless, as the perforation diameter decreases, Figure 15 shows that the difference in the fracture propagation between the cases considering perforation erosion and the cases without regard to perforation erosion increases. The difference in the central fracture length is 11 m between the two cases when the perforation diameter is 8 mm, while the difference decreases to 1 m when the perforation diameter is 11 mm, which indicates that perforation erosion has a more pronounced impact when the perforation diameter is small. This is because the mean velocity at the perforation increases when the perforation diameter decreases, which results in greater variation in the diameter and discharge coefficient of the perforations. Figure 16 illustrates that the discharge coefficient of the perforations changes the most when their diameter is 8 mm. Therefore, the effects of erosion on multi-fracture propagation are enhanced as the perforation diameter decreases because the perforation erosion is more severe when the perforation diameter is small.

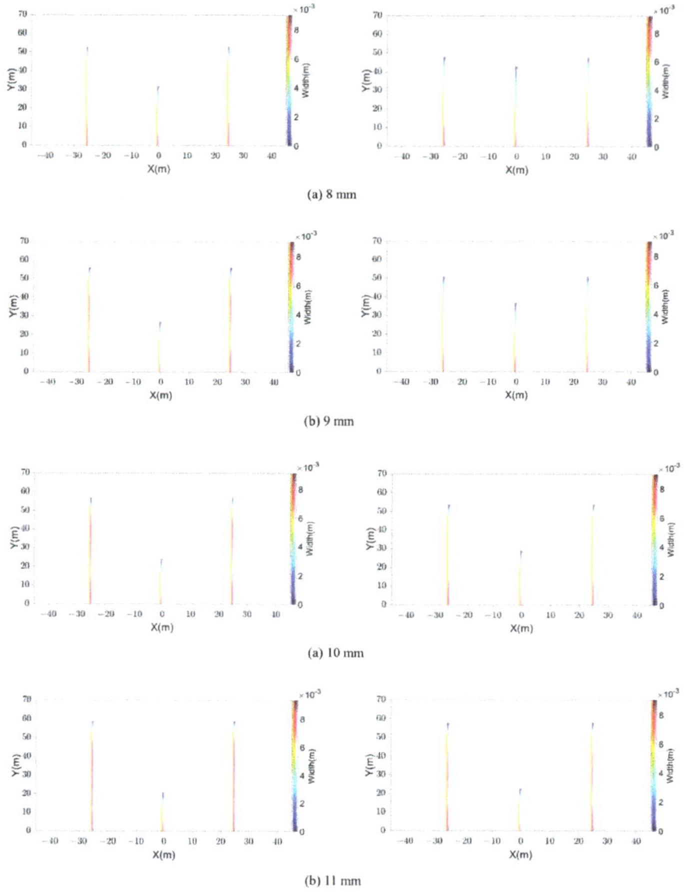

Figure 13. Fracture propagation under different perforation diameters (left: erosion; right: no erosion).

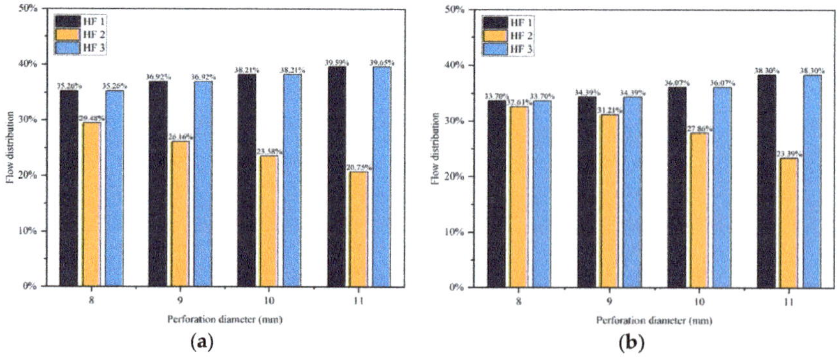

Figure 14. The flow distribution between fractures under different perforation diameters: (**a**) erosion; (**b**) no erosion.

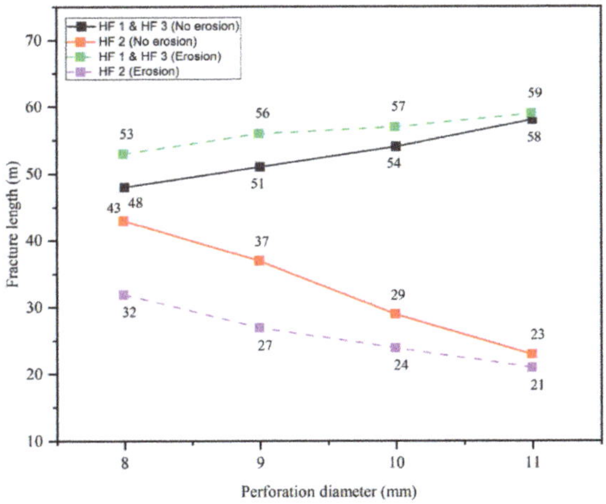

Figure 15. The length of each fracture under different perforation diameters.

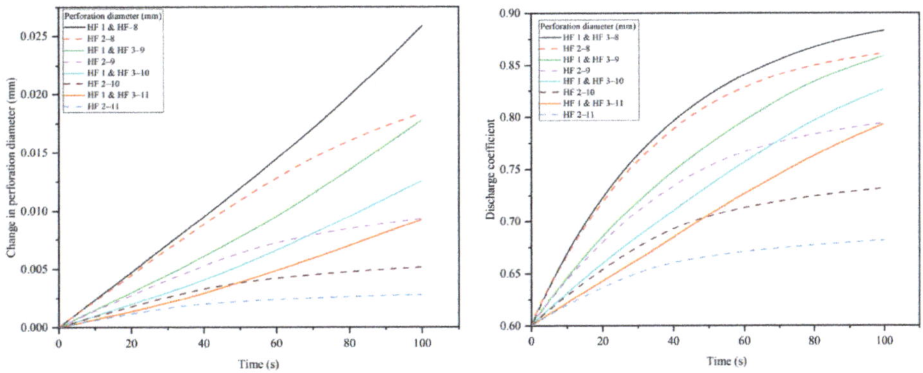

(a) (b)

Figure 16. The change in perforation diameter and discharge coefficient under different perforation diameters: (**a**) perforation diameter; (**b**) discharge coefficient.

As discussed above, the effects of erosion on multi-fracture propagation are enhanced as the perforation diameter decreases. Although the effects of erosion on multi-fracture propagation can be mitigated by increasing the perforation diameter, the increase in the perforation diameter decreases the perforation friction and aggravates uneven fracture propagation. Thus, to overcome the effect of the perforation erosion, a smaller perforation diameter is recommended to ensure the initial perforation friction is much higher than the perforation friction after perforation erosion.

4.4. The Effects of the Perforation Number

The perforation number is one of the crucial parameters in the optimization of multi-cluster fracturing design. Limited-entry fracturing is achieved by adjusting the perforation number and perforation diameter. In this section, four cases are set to investigate the multi-fracture propagation under different perforation numbers (8, 10, 12, and 14).

Figure 17 illustrates the morphology of the fractures under different perforation numbers. It can be found that reducing the perforation number is beneficial for promoting uniform fracture propagation. And Figure 18 shows the flow distribution between the fractures becomes uniform as the perforation number decreases, which is beneficial for promoting uniform fracture propagation. Nevertheless, perforation erosion diminishes the capacity of the perforation number to facilitate uniform propagation of the fractures. When taking into account perforation erosion, it is observed that the length of the side fracture is longer and the length of central fracture is shorter, which increases the non-uniformity of the fracture propagation, and the erosion effects become more pronounced as the perforation number decreases. Figure 19 illustrates a 3 m disparity in the central fracture length between the cases accounting for perforation erosion and those disregarding it with a perforation number of 14, and the difference increases to 10 m when the perforation number is 8. This is because the perforation erosion is more severe with a smaller perforation number, which aggravates uneven propagation of the fractures. Figure 20 illustrates the variation in the diameter and discharge coefficient of the perforations under different perforation numbers. When the perforation number is 14, the variation in the diameter and discharge coefficient of the perforations is small. And when the perforation number is 8, the diameter and the discharge coefficient of the perforations change the most, which suggests that a smaller perforation number corresponds to more severe perforation erosion.

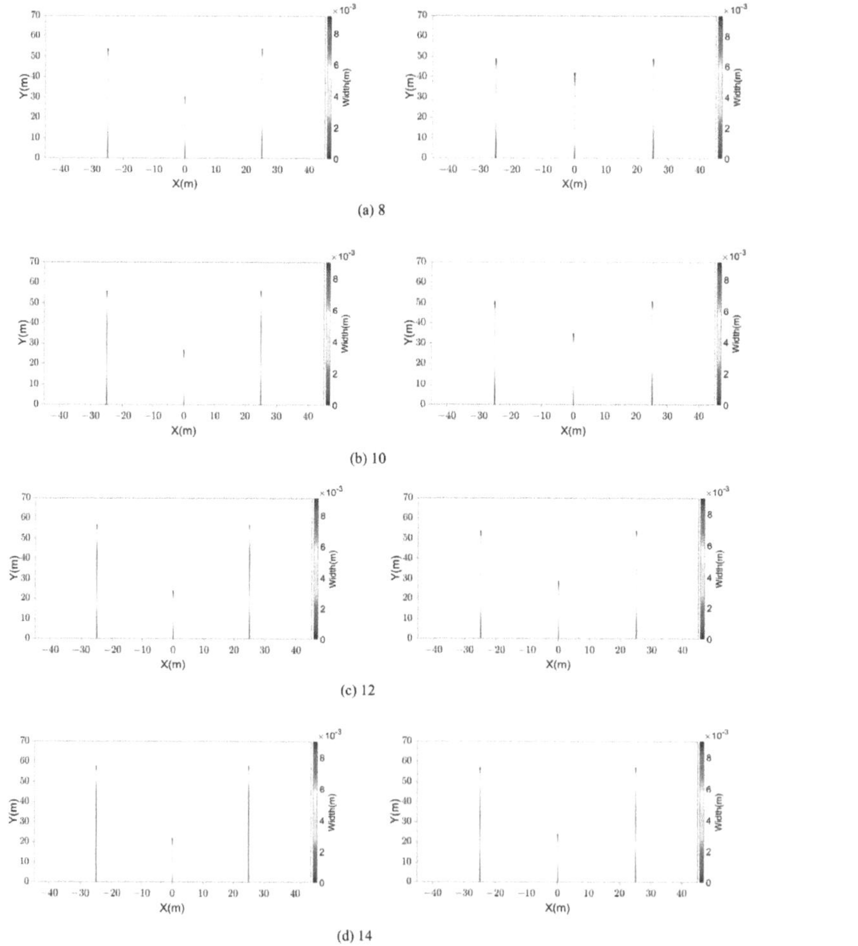

Figure 17. Fracture propagation under different perforation numbers (left: erosion; right: no erosion).

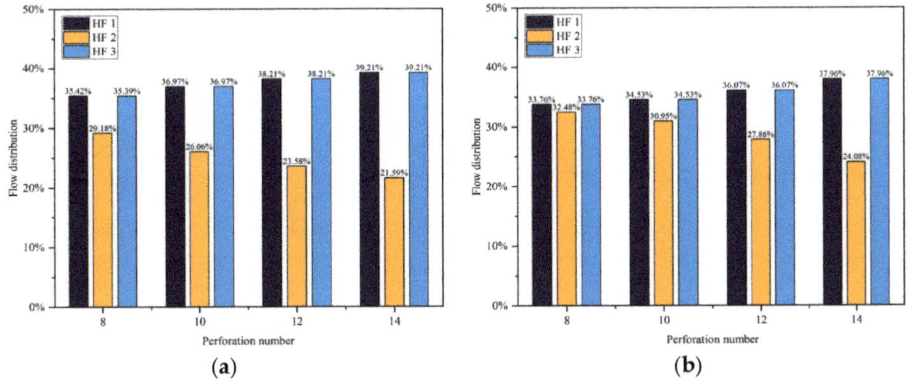

Figure 18. The flow distribution between fractures under different perforation numbers: (**a**) erosion; (**b**) no erosion.

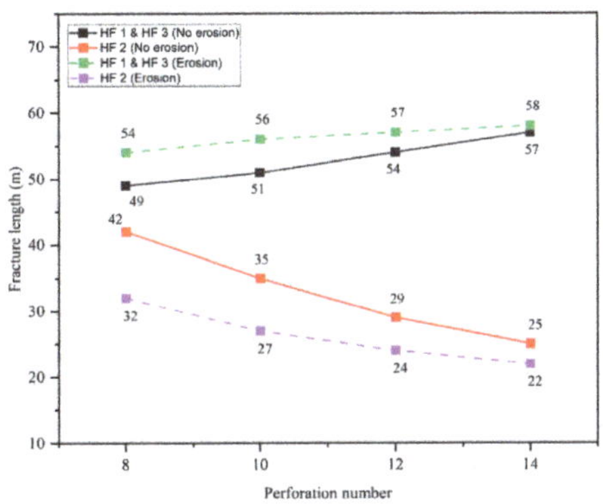

Figure 19. The length of each fracture under different perforation numbers.

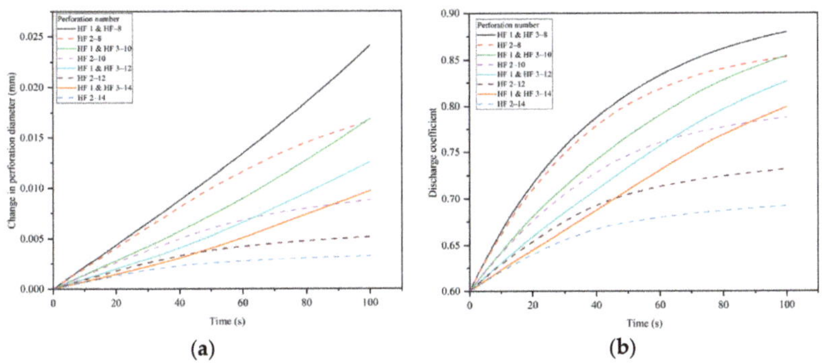

Figure 20. The change in perforation diameter and discharge coefficient under different perforation numbers: (**a**) perforation diameter; (**b**) discharge coefficient.

In conclusion, the effects of erosion on multi-fracture propagation are aggravated as the perforation number decreases. And when the perforation number increases, the erosion of the perforations is mitigated. Nevertheless, although the effects of erosion on multi-fracture propagation are more severe when the perforation number decreases, reducing the perforation number is still conductive to promoting uniform fracture propagation. Thus, a smaller perforation number can be considered to overcome the effect of perforation erosion and ensure uniform fracture propagation.

4.5. The Effects of Cluster Spacing

The cluster spacing is closely associated with the stress interference between fractures and has a significant influence on multi-fracture propagation. In this section, four cases are considered to study the multi-fracture propagation under different cluster spacings (20 m, 25 m, 30 m, and 35 m).

Figure 21 shows the fracture morphology with four different cluster spacings. The impact of stress shadow decreases as the cluster spacing increases. Hence, the uniformity of the fracture propagation increases as the cluster spacing increases. However, an increase in the cluster spacing enhances the impact of perforation erosion. Figure 22 shows that the difference in the length of the central fracture between the cases considering perforation erosion and those disregarding perforation erosion is 2 m when the cluster spacing is 20 m, and the difference increases to 11 m when it is 35 m. This is because the stress shadow effect is strong with a smaller cluster spacing, and a higher perforation friction is needed to ensure uniform fracture propagation. Despite the perforation erosion resulting in a reduction in the perforation friction, the multi-fracture propagation is less affected. As the cluster spacing increases, the effect of stress shadow is weakened, and the decrease in the perforation friction can significantly affect the multi-fracture propagation. Hence, although Figure 23 shows that the difference in the perforation erosion under different cluster spacings is small, the effect of the perforation erosion is enhanced as the cluster spacing increases.

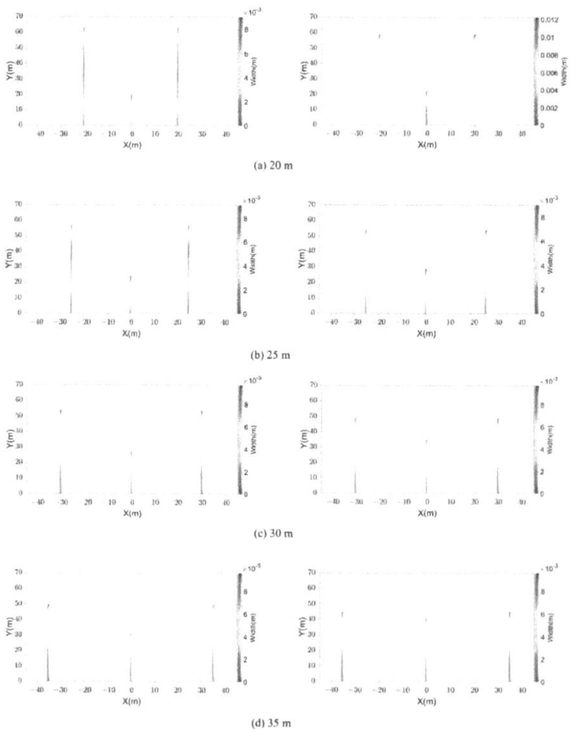

Figure 21. Fracture propagation with different cluster spacings (left: erosion; right: no erosion).

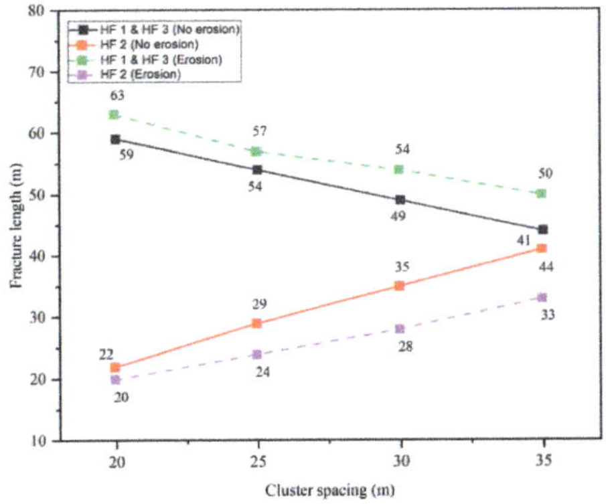

Figure 22. The length of each fracture under different cluster spacings.

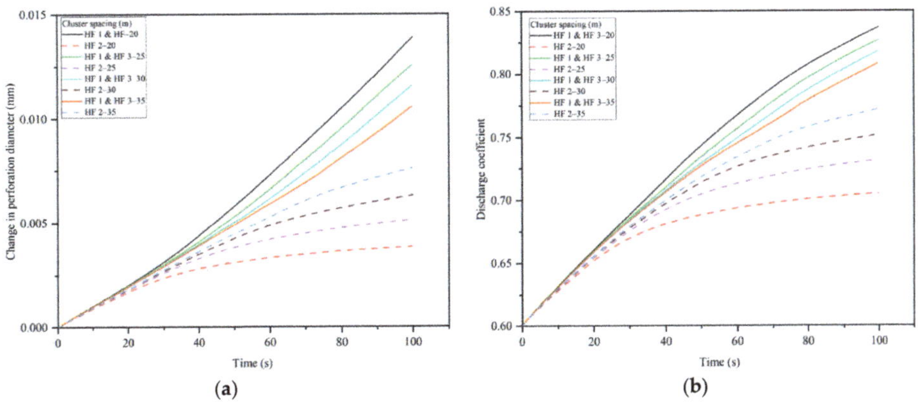

Figure 23. The change in perforation diameter and discharge coefficient under different cluster spacings: (**a**) perforation diameter; (**b**) discharge coefficient.

Furthermore, the increase in the cluster spacing reduces the difference in the perforation erosion between the central fractures and the side fractures, as shown in Figure 23. The increase in the cluster spacing contributes to a more uniform flow distribution. As depicted in Figure 24, the fluid distributed to the central fracture increases while the fluid distributed to the side fractures decreases as the cluster spacing increases, which suggests that the injection rate in the side fractures decreases while the injection rate in the central fracture increases. As discussed in Section 4.2, the erosion of the perforations is more severe at an increasing injection rate. Therefore, the erosion of the side fractures is reduced while the erosion of the perforations in the central fracture is intensified, and the difference in the perforation erosion between the central fractures and the side fractures is reduced as the cluster spacing increases.

In conclusion, the cluster spacing has less influence on the perforation erosion. However, the stress shadow effect between fractures can diminish the effect of perforation erosion. The effects of erosion on multi-fracture propagation become more pronounced as the cluster spacing is increased. Nevertheless, increasing the cluster spacing can still contribute to uniform fracture propagation. Thus, larger cluster spacing can be considered to promote uniform fracture propagation.

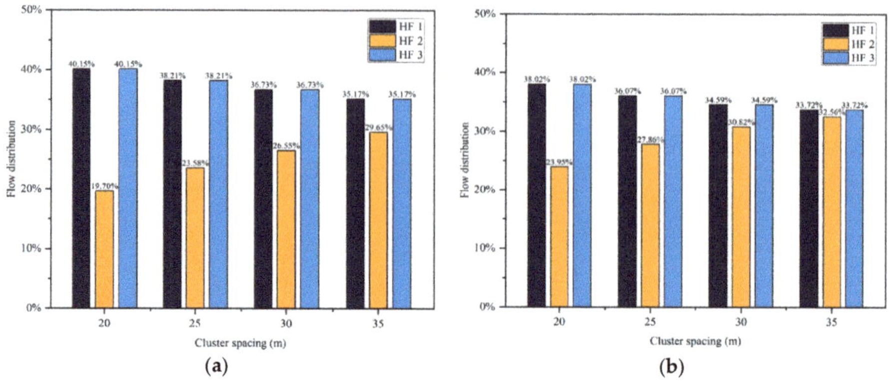

Figure 24. The flow distribution between fractures with different cluster spacings: (**a**) erosion; (**b**) no erosion.

5. Conclusions

In this study, a 2D hydraulic fracturing model is developed to study the multi-fracture propagation considering perforation erosion under different parameters. The study can provide recommendations for the optimization of multi-stage fracturing design and is helpful for enhancing shale reservoir stimulation. The following conclusions are obtained:

1. Perforation erosion increases the discharge coefficient and perforation diameter, which results in a reduction in the perforation friction and aggravates uneven fracture propagation.
2. For a certain volume of fracturing fluid, although the injection time is longer at a lower injection rate, the effects of erosion on multi-fracture propagation are more severe at a high injection rate. Nevertheless, an increasing injection rate is still beneficial for promoting uniform fracture propagation, and a higher injection rate is recommended.
3. When the perforation number and perforation diameter decrease, the erosion of the perforations is aggravated, which leads to a sharp decrease in the perforation friction and aggravates the effects of erosion on the multi-fracture propagation. To overcome the effect of perforation erosion, a smaller perforation diameter and perforation number are recommended to ensure the initial perforation friction is much higher than the perforation friction after perforation erosion.
4. The cluster spacing has less effect on perforation erosion. And stress shadow can diminish the effects of erosion on multi-fracture propagation, which indicates that the effects of erosion on multi-fracture propagation are enhanced as the cluster spacing is increased. Nevertheless, increasing the cluster spacing is still conducive to promoting uniform fracture propagation, and larger cluster spacing can be considered to promote uniform fracture propagation.

Author Contributions: Methodology: L.X.; software: L.T.; data curation: L.T.; writing—original draft preparation, review and editing: B.H. and Y.Z. All authors have read and agreed to the published version of the manuscript.

Funding: This work was financially supported by the National Natural Science Foundation (No. 11872258).

Data Availability Statement: The data presented in this study are available on request from the corresponding author. The data are not publicly available due to future research.

Conflicts of Interest: The authors declare that they have no known competing financial interests or personal relationships that could have appeared to influence the work reported in this paper.

References

1. Yuan, J.; Luo, D.; Feng, L. A review of the technical and economic evaluation techniques for shale gas development. *Appl. Energy* **2015**, *148*, 49–65. [CrossRef]
2. U.S. EIA. *Technically Recoverable Shale Oil and Shale Gas Resources: An Assessment of 137 Shale Formation: 41 Countries outside the United States*; U.S. EIA: Washington, DC, USA, 2013.
3. Niu, W.; Lu, J.; Sun, Y.; Liu, H.; Cao, X.; Zhan, H.; Zhang, J. A review of the application of data-driven technology in shale gas production evaluation. *Energy Rep.* **2023**, *10*, 213–227. [CrossRef]
4. Li, H. Coordinated development of shale gas benefit exploitation and ecological environmental conservation in China: A mini review. *Front. Ecol. Evol.* **2023**, *11*. [CrossRef]
5. Liu, J.; Bai, X.; Elsworth, D. Evolution of pore systems in low-maturity oil shales during thermal upgrading—Quantified by dynamic SEM and machine learning. *Pet. Sci.* **2023**, *in press*. [CrossRef]
6. Xu, Y.; Lun, Z.; Pan, Z.; Wang, H.; Zhou, X.; Zhao, C.; Zhang, D. Occurrence space and state of shale oil: A review. *J. Pet. Sci. Eng.* **2022**, *211*, 110183. [CrossRef]
7. Tang, H.; Winterfeld, P.H.; Wu, Y.-S.; Huang, Z.-q.; Di, Y.; Pan, Z.; Zhang, J. Integrated simulation of multi-stage hydraulic fracturing in unconventional reservoirs. *J. Nat. Gas Sci. Eng.* **2016**, *36*, 875–892. [CrossRef]
8. Sobhaniaragh, B.; Mansur, W.J.; Peters, F.C. Three-dimensional investigation of multiple stage hydraulic fracturing in unconventional reservoirs. *J. Pet. Sci. Eng.* **2016**, *146*, 1063–1078. [CrossRef]
9. Zhao, J.; Zhao, J.; Hu, Y.; Zhang, S.; Huang, T.; Liu, X. Numerical simulation of multistage fracturing optimization and application in coalbed methane horizontal wells. *Eng. Fract. Mech.* **2020**, *223*, 106738. [CrossRef]
10. Modeland, N.; Buller, D.; Chong, K.K. Statistical Analysis of the Effect of Completion Methodology on Production in the Haynesville Shale. In Proceedings of the North American Unconventional Gas Conference and Exhibition, The Woodlands, TX, USA, 12–16 June 2011; p. SPE-144120-MS.
11. Cheng, Y. Impacts of the Number of Perforation Clusters and Cluster Spacing on Production Performance of Horizontal Shale-Gas Wells. *SPE Reserv. Eval. Eng.* **2012**, *15*, 31–40. [CrossRef]
12. Wu, Q.; Xu, Y.; Wang, X.; Wang, T.; Zhang, S. Volume fracturing technology of unconventional reservoirs: Connotation, design optimization and implementation. *Pet. Explor. Dev.* **2012**, *39*, 377–384. [CrossRef]
13. Miller, C.; Waters, G.; Rylander, E. Evaluation of Production Log Data from Horizontal Wells Drilled in Organic Shales. In Proceedings of the North American Unconventional Gas Conference and Exhibition, The Woodlands, TX, USA, 12–16 June 2011.
14. Ugueto, C.G.A.; Huckabee, P.T.; Molenaar, M.M.; Wyker, B.; Somanchi, K. Perforation Cluster Efficiency of Cemented Plug and Perf Limited Entry Completions; Insights from Fiber Optics Diagnostics. In Proceedings of the SPE Hydraulic Fracturing Technology Conference, The Woodlands, TX, USA, 9–11 February 2016.
15. Rainbolt, M.F.; Esco, J. Frac Hit Induced Production Losses: Evaluating Root Causes, Damage Location, Possible Prevention Methods and Success of Remediation Treatments, Part II. In Proceedings of the SPE Hydraulic Fracturing Technology Conference and Exhibition, Muscat, Oman, 16–18 October 2018; p. D021S006R004.
16. Li, Y.; Deng, J.; Liu, W.; Yan, W.; Feng, Y.; Cao, W.; Wang, P.; Hou, Y. Numerical simulation of limited-entry multi-cluster fracturing in horizontal well. *J. Pet. Sci. Eng.* **2017**, *152*, 443–455. [CrossRef]
17. Zeng, Q.; Liu, Z.; Wang, T.; Gao, Y.; Zhuang, Z. Fully coupled simulation of multiple hydraulic fractures to propagate simultaneously from a perforated horizontal wellbore. *Comput. Mech.* **2018**, *61*, 137–155. [CrossRef]
18. Zheng, P.; Xia, Y.; Yao, T.; Jiang, X.; Xiao, P.; He, Z.; Zhou, D. Formation mechanisms of hydraulic fracture network based on fracture interaction. *Energy* **2022**, *243*, 123057. [CrossRef]
19. ElGibaly, A.; Abdalla Osman, M. Perforation friction modeling in limited entry fracturing using artificial neural network. *Egypt. J. Pet.* **2019**, *28*, 297–305. [CrossRef]
20. Zhou, T.; Chen, M.; Zhang, S.; Li, Y.; Li, F.; Zhang, C. Simulation of fracture propagation and optimization of ball-sealer in-stage diversion under the effect of heterogeneous stress field. *Nat. Gas Ind. B* **2020**, *7*, 523–532. [CrossRef]
21. Lim, P.; Goddard, P.; Sink, J.; Abou-Sayed, I. Hydraulic Fracturing: A Marcellus Case Study of an Engineered Staging Completion Based on Rock Properties. In Proceedings of the SPE/CSUR Unconventional Resources Conference—Canada, Calgary, AB, Canada, 30 September–2 October 2014; p. D011S003R006.
22. Wu, K.; Olson, J.; Balhoff, M.T.; Yu, W. Numerical Analysis for Promoting Uniform Development of Simultaneous Multiple-Fracture Propagation in Horizontal Wells. *SPE Prod. Oper.* **2016**, *32*, 41–50.
23. Peirce, A.P.P.; Bunger, A.P.P. Interference Fracturing: Nonuniform Distributions of Perforation Clusters That Promote Simultaneous Growth of Multiple Hydraulic Fractures. *SPE J.* **2014**, *20*, 384–395. [CrossRef]
24. Chen, M.; Zhang, S.; Ma, X.; Zhou, T.; Zou, Y. A semi-analytical model for predicting fluid partitioning among multiple hydraulic fractures from a horizontal well. *J. Pet. Sci. Eng.* **2018**, *171*, 1041–1051. [CrossRef]
25. Zhang, F.; Wang, X.; Tang, M.; Du, X.; Xu, C.; Tang, J.; Damjanac, B. Numerical Investigation on Hydraulic Fracturing of Extreme Limited Entry Perforating in Plug-and-Perforation Completion of Shale Oil Reservoir in Changqing Oilfield, China. *Rock Mech. Rock Eng.* **2021**, *54*, 2925–2941. [CrossRef]
26. Li, M.; Lv, W.; Liu, J.; Sun, Z.; Zhou, F.; Wang, B. Effect of perforation friction on 3D In-stage multiple fracture propagation: A numerical study. *Eng. Fract. Mech.* **2022**, *267*, 108415. [CrossRef]

27. Lecampion, B.; Desroches, J.; Weng, X.; Burghardt, J.; Brown, J.E.E. Can We Engineer Better Multistage Horizontal Completions? Evidence of the Importance of Near-wellbore Fracture Geometry from Theory, Lab and Field Experiments. In Proceedings of the SPE Hydraulic Fracturing Technology Conference, The Woodlands, TX, USA, 3–5 February 2015; p. D021S003R001.
28. Somanchi, K.; O'Brien, C.; Huckabee, P.; Ugueto, G. Insights and Observations into Limited Entry Perforation Dynamics from Fiber-Optic Diagnostics. In Proceedings of the SPE/AAPG/SEG Unconventional Resources Technology Conference, San Antonio, TX, USA, 1–3 August 2016; p. URTEC-2458389-MS.
29. Roberts, G.; Whittaker, J.; McDonald, J.; Paxson, T. Proppant Distribution Observations from 20,000+ Perforation Erosion Measurements. In Proceedings of the SPE Hydraulic Fracturing Technology Conference and Exhibition, The Woodlands, TX, USA, 4–6 February 2020; p. D011S002R002.
30. Crump, J.B.; Conway, M.W. Effects of Perforation-Entry Friction on Bottomhole Treating Analysis. *J. Pet. Technol.* **1988**, *40*, 1041–1048. [CrossRef]
31. Roberts, G.; Lilly, T.B.; Tymons, T.R. Improved Well Stimulation through the Application of Downhole Video Analytics. In Proceedings of the SPE Hydraulic Fracturing Technology Conference and Exhibition, The Woodlands, TX, USA, 23–25 January 2018; p. D011S001R005.
32. Cramer, D.; Friehauf, K.; Roberts, G.; Whittaker, J. Integrating Distributed Acoustic Sensing, Treatment-Pressure Analysis, and Video-Based Perforation Imaging to Evaluate Limited-Entry-Treatment Effectiveness. *SPE Prod. Oper.* **2020**, *35*, 0730–0755. [CrossRef]
33. Guo, J.; Zhao, X.; Zhu, H.; Zhang, X.; Pan, R. Numerical simulation of interaction of hydraulic fracture and natural fracture based on the cohesive zone finite element method. *J. Nat. Gas Sci. Eng.* **2015**, *25*, 180–188. [CrossRef]
34. Zhao, P.; Xie, L.; Ge, Q.; Zhang, Y.; Liu, J.; He, B. Numerical study of the effect of natural fractures on shale hydraulic fracturing based on the continuum approach. *J. Pet. Sci. Eng.* **2020**, *189*, 107038. [CrossRef]
35. Dong, Y.; Tian, W.; Li, P.; Zeng, B.; Lu, D. Numerical investigation of complex hydraulic fracture network in naturally fractured reservoirs based on the XFEM. *J. Nat. Gas Sci. Eng.* **2021**, *96*, 104272. [CrossRef]
36. Li, Z.-Q.; Li, X.-L.; Yu, J.-B.; Cao, W.-D.; Liu, Z.-F.; Wang, M.; Liu, Z.-F.; Wang, X.-H. Influence of existing natural fractures and beddings on the formation of fracture network during hydraulic fracturing based on the extended finite element method. *Geomech. Geophys. Geo-Energy Geo-Resour.* **2020**, *6*, 58. [CrossRef]
37. Bažant, Z.P.; Oh, B.H. Crack band theory for fracture of concrete. *Matériaux Et Constr.* **1983**, *16*, 155–177. [CrossRef]
38. Zienkiewicz, O.C.; Shiomi, T. Dynamic behaviour of saturated porous media; The generalized Biot formulation and its numerical solution. *Int. J. Numer. Anal. Methods Geomech.* **1984**, *8*, 71–96. [CrossRef]
39. Li, S.; Li, X.; Zhang, D. A fully coupled thermo-hydro-mechanical, three-dimensional model for hydraulic stimulation treatments. *J. Nat. Gas Sci. Eng.* **2016**, *34*, 64–84. [CrossRef]
40. Zhou, L.; Hou, M.Z. A new numerical 3D-model for simulation of hydraulic fracturing in consideration of hydro-mechanical coupling effects. *Int. J. Rock Mech. Min. Sci.* **2013**, *60*, 370–380. [CrossRef]
41. Roostaei, M.; Taghipoor, S.; Nouri, A.; Fattahpour, V.; Chan, D. Smeared modeling of hydraulic fracture using partially coupled reservoir and geomechanics simulators. *Int. J. Rock Mech. Min. Sci.* **2019**, *113*, 99–111. [CrossRef]
42. Chen, X.; Zhao, J.; Li, Y.; Yan, W.; Zhang, X. Numerical Simulation of Simultaneous Hydraulic Fracture Growth within a Rock Layer: Implications for Stimulation of Low-Permeability Reservoirs. *J. Geophys. Res.-Solid Earth* **2019**, *124*, 13227–13249. [CrossRef]
43. Lecampion, B.; Desroches, J. Simultaneous initiation and growth of multiple radial hydraulic fractures from a horizontal wellbore. *J. Mech. Phys. Solids* **2015**, *82*, 235–258. [CrossRef]
44. Long, G.; Liu, S.; Xu, G.; Wong, S.-W. Modeling of Perforation Erosion for Hydraulic Fracturing Applications. In Proceedings of the SPE Annual Technical Conference and Exhibition, Houston, TX, USA, 28–30 September 2015; p. D021S015R008.
45. Valko, P.P.; Economides, M.J. *Hydraulic Fracture Mechanics*; Wiley: Chichester, UK, 1995.
46. Xing, P.; Yoshioka, K.; Adachi, J.; El-Fayoumi, A.; Bunger, A.P. Laboratory measurement of tip and global behavior for zero-toughness hydraulic fractures with circular and blade-shaped (PKN) geometry. *J. Mech. Phys. Solids* **2017**, *104*, 172–186. [CrossRef]
47. Harris, P.C.; Pippin, P.M. High-Rate Foam Fracturing: Fluid Friction and Perforation Erosion. *SPE Prod. Facil.* **2000**, *15*, 27–32. [CrossRef]
48. Shin, D.H.; Sharma, M.M. Factors Controlling the Simultaneous Propagation of Multiple Competing Fractures in a Horizontal Well. In Proceedings of the SPE Hydraulic Fracturing Technology Conference, The Woodlands, TX, USA, 4–6 February 2014; p. SPE-168599-MS.
49. Yang, C.-X.; Yi, L.-P.; Yang, Z.-Z.; Li, X.-G. Numerical investigation of the fracture network morphology in multi-cluster hydraulic fracturing of horizontal wells: A DDM-FVM study. *J. Pet. Sci. Eng.* **2022**, *215*, 110723. [CrossRef]

energies

MDPI

Article

Numerical Simulation of the Simultaneous Development of Multiple Fractures in Horizontal Wells Based on the Extended Finite Element Method

Enshun Ping [1], Peng Zhao [2,*], Haiyan Zhu [3,4], Yuzhong Wang [1], Zixi Jiao [3,4], Qingjie Zhao [1] and Gan Feng [5,6]

1 Downhole Technology Service Company, CNPC Bohai Drilling Engineering Company Limited, Tianjin 300457, China; pingenshun@163.com (E.P.); wangyuzhong@cnpc.com.cn (Y.W.); zhaoqingjie@cnpc.com.cn (Q.Z.)
2 State Key Laboratory of Geohazard Prevention and Geoenvironment Protection, Chengdu University of Technology, Chengdu 610059, China
3 State Key Laboratory of Oil and Gas Reservoir Geology and Exploitation, Chengdu University of Technology, Chengdu 610059, China; zhuhaiyan040129@163.com (H.Z.); jiaozixi414@163.com (Z.J.)
4 College of Energy, Chengdu University of Technology, Chengdu 610059, China
5 State Key Laboratory of Hydraulics and Mountain River Engineering, Sichuan University, Chengdu 610065, China; fenggan@whu.edu.cn
6 College of Water Resource & Hydropower, Sichuan University, Chengdu 610065, China
* Correspondence: scu_zhaopeng@163.com

Abstract: Simultaneous multiple-fracture treatments in horizontal wellbores have become an essential technology for the economic development of shale gas reservoirs. During hydraulic fracturing, fracture initiation and propagation always induce additional stresses on the surrounding rock. When multiple fractures develop simultaneously, the development of some fractures is limited due to the stress-shadow effect. An in-depth understanding of the multiple-fracture propagation mechanism as reflected by fracture morphology and flow rate distribution can help to set reasonable operation parameters for improving the uniformity of multiple fractures and forming a complex fracture network according to the different in situ stress conditions in a reservoir to increase the shale gas recovery and reduce the cost. In this study, a two-dimensional (2D) fracture propagation model was developed based on the extended finite element method (XFEM). Then, the influences of various factors, including geological and operational factors, on the development of multiple fractures were studied. The results of numerical simulations showed that increasing the cluster spacing or injecting fracturing fluid with lower viscosity can help reduce the stress-shadow effect. In the case of smaller in situ stress differences, the deflection of the fractures was larger due to the stress-shadow effect. As the stress difference increased, the direction of the propagation of the fracture was gradually biased towards the direction of maximum horizontal stress. In addition, the injection rate had some effects on the fracture morphology and flow rate distribution. However, as the injection rate increased, the dominant fracture developed rapidly, and the fracture length significantly increased.

Keywords: XFEM; stress-shadow effect; competition propagation of fractures; in situ stress difference; injection rate

1. Introduction

The development of gas in low-porosity/permeability reservoirs such as shale gas reservoirs is a solution to the growing demand for energy resources [1,2]. Meanwhile, shale gas, as a clean energy source like geothermal energy, has low-carbon and green characteristics and has taken on an important role in global clean energy [3]. Shale gas reservoirs have poorer physical properties and more complex in situ stress conditions than those of conventional reservoirs, which has led to the use of a number of stimulation techniques during shale gas development [4,5]. Over the last decades, reservoir stimulation techniques combining horizontal well technology with multi-stage hydraulic fracturing have played an important role in shale gas development, enlarging the drainage area by arranging multiple perforation clusters within a stage, thereby increasing recovery and reducing operating costs. Although this technique has been applied more often in practical engineering, the mechanism of the influences of different geological and operational parameters on the simultaneous development of multiple fractures is not completely clear, and in-depth studies are needed [6–8].

During the fracturing process, it is desired that multiple fractures develop and propagate uniformly with minimal resistance. However, the results of mine experiments and laboratory tests have shown that even in homogeneous formations, hydraulic fractures have characteristics such as deflection and offsets that are opposite to the two-wing and planar characteristics [9]. The morphology of fractures is decided by various factors, such as the injection rate and in situ stress conditions. Warpinski et al. [10] conducted mine experiments in ash-fall tuffs with low permeability and showed that in situ stress affects the propagation mode of hydraulic fractures. Zeng et al. [11] carried out experiments on the development of hydraulic fractures using cement samples. When the well pipe position was changed, the complexity of the hydraulic fracture morphology was different due to the stress differences. Some researchers have investigated the effects of operation factors on hydraulic fracture morphology [12,13]. The experimental results showed that the complexity of the fractures and the probability of fracture deflection increased with the increase in the viscosity of the injected fluid. In addition, the higher the injection rate, the lower the complexity of the fracture and the greater the likelihood of planar fracture.

Meanwhile, when multiple fractures develop simultaneously, the induced stresses generated between the fractures will change the in situ stress distribution, causing the hydraulic fracture's propagation direction to deviate from the direction of the maximum principal stress, resulting in a complex fracture morphology. This phenomenon is known as the stress-shadow effect, and it usually results in the suppression of the center fracture, while the fractures on either side are overextended and show varying degrees of deflection [14]. Qu et al. [15] found that when multiple fractures occur simultaneously, the induced stresses arise at the fracture tip, and the stress field around each fracture is different. Fractures in which stress is concentrated at the tip require more energy to promote fracture propagation. Fallahzadeh et al. [16] set up two perforation clusters in mortar samples. Once two fractures were initiated, they propagated in opposite directions, and fracture tortuosity occurred in an area near the wellbore. The proportion of fluid in a fracture after applying fracturing fluid determines the morphology and development of the fracture. In general, fracturing fluids tend to flow into fractures with minimal resistance. The stress-shadow effect generates compressive stress on fracture surfaces, resulting in a different resistance to flow within each fracture. During the process of the flow rate distribution, more fluid enters fractures with lower compressive stress [17–19]. As a result, each fracture develops differently. Miller et al. [20] concluded from production logging data from a large number of horizontal shale wells that most shale hydrocarbons came from a small number of perforation clusters. Ketterij et al. [21] set up multiple clusters and carried out hydraulic fracturing experiments. Their experiment showed that the amount of fluid entering each fracture varied. While it has been recognized that stress-shadow effects can directly affect the complexity of fracture networks, more work needs to be performed to explore the specific roles of different parameters in fracture morphology and flow rate distributions.

Due to the lack of effective diagnostic techniques for fractures, complex fracture geometries caused by the development of multiple fractures are not accurately characterized during mine experiments [8,22]. In laboratory experiments, it is difficult to ensure that conditions are completely consistent, especially with true samples, and the scale of simulations is limited. With the improvement of various numerical methods and computer efficiency, numerical simulation methods have gradually been applied to larger-scale fracture studies in shale reservoirs [23–25]. In previous studies, researchers used the finite element method

(FEM) to simulate fracture generation processes. In this method, the cohesive zone model (CZM), which is commonly used to describe fracture initiation and propagation, obeys a law called "traction–separation" [26–32]. Guo et al. [26] established a three-dimensional model to study the simultaneous generation of multiple fractures in shale reservoirs with a thick pay zone. Shin et al. [27] discussed the effects of the number of perforation clusters, injected fluid viscosity, and elastic modulus on the development state of multiple fractures that were initiated simultaneously. The results showed that geological and operational factors have a significant impact on the morphology of the fractures and the distribution of fluids in them. Recently, the extended finite element method (XFEM) has been applied in the field of computational mechanics [33–36]. This method represents fractures or material interfaces in a computational region by means of discontinuous form functions that are independent of the element network. The advantage of XFEM is that there is no need to update the finite element mesh to trace the location of a fracture. In addition, fractures can propagate along arbitrary paths. Compared to CZM, XFEM is better suited for showing the complexities of fracture development, such as with the deflection angle [37,38]. Currently, researchers mainly discuss the patterns of interactions between induced fractures and pre-existing fractures through XFEM. More work needs to be performed to study the simultaneous propagation of multiple fractures by considering more factors [39–41]. The complexity of a fracture network depends on the morphology of each fracture. Importantly, the stress-shadow effect plays a crucial role in the degree of uniform planar propagation of multiple fractures. Therefore, the influences of various factors (i.e., geological and operational factors) on the stress-shadow effect should be discussed. Reduction in the negative impact of the stress-shadow effect and the formation of a complex fracturing network are essential for improving the efficiency of shale gas extraction.

In this study, a two-dimensional hydraulic fracture propagation model based on the XFEM was established. Five fractures were embedded in a coupled hydro-mechanic model. The competing propagation of fractures was simulated under different conditions. Several factors were considered, including in situ stress differences, cluster space, fracturing fluid viscosity, and injection rate. The effects of geological and operational factors on the fracture network's morphology are discussed. This study contributes to the understanding of the mechanism of the stress-shadow effect on multiple-fracture propagation. These conclusions help in recognizing the key factors in controlling the uniform growth of multiple fractures, which can help to improve the uniformity of multiple fractures and form a complex fracture network, which will be conducive to improving shale gas recovery and reducing costs.

2. Governing Equations

2.1. Model Assumptions

The hydraulic fracturing process involves the complex dynamic coupling of fluids and solids. In order to comprehensively study the process of the simultaneous propagation of multiple fractures with reduced computational burden, some assumptions need to be considered [42]:

(1) The rock matrix material is considered to be homogeneous and elastic;
(2) The injected fluid is considered incompressible;
(3) The proppant transport within multiple fractures is not taken into account;
(4) The influence of the temperature field is not considered in the simulation.

2.2. The Flow Mode of the Fluid

During fracture generation, the continuous injection of a fluid triggers a fracture to propagate. To describe this complex phenomenon, we need two sets of laws [42]:

(1) In the fluid section, we need to describe the fluid flow process and the fluid filtration loss process within the fracture;
(2) In the soil section, we need some initiation and propagation criteria to describe the whole process from the deformation to the destruction of rocks.

The directions of fluid flow within a fracture include the tangential and normal directions (Figure 1). Along the tangential direction, fluid flows into the fracture, while along the other direction, fluid is lost by filtering from the fracture into the rock mass. The equation of fluid flow in the tangential direction was given by Reynold [26]:

$$q = -\frac{w^3}{12\mu}\nabla p \tag{1}$$

where q represents the flow rate along with tangential direction, w represents the width of the fracture, μ represents the viscosity of the injected fluid, and ∇p represents the pressure variation inside the fracture.

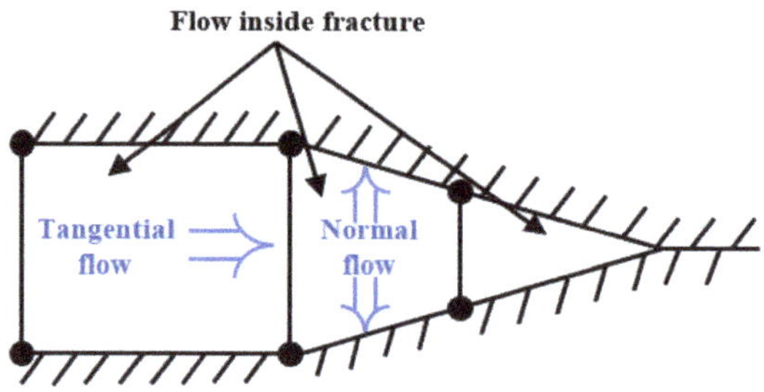

Figure 1. Fluid flow in the fracture.

The equation of the fluid flow along the normal direction can be written as follows [26]:

$$\begin{cases} q_t = c_t(p_i - p_t) \\ q_b = c_b(p_i - p_b) \end{cases} \tag{2}$$

where q_t represents the flow rate at the upper fracture surface, q_b represents the flow rate at the lower fracture surface, c_t represents the coefficient of filtration loss at the upper fracture surface, c_b represents the coefficient of filtration loss at the lower fracture surface, p_i represents the pressure at the mid-face, p_t represents the pore pressure at the upper face, and p_b represents the pore pressure at the bottom face.

2.3. The Formula for Effective Stress

For rock with a certain porosity, the stress can be expressed as follows [26]:

$$\sigma_{total} = \sigma' + \alpha p_w \delta \tag{3}$$

where σ_{total} represents all stress in the rock, σ' represents the effective stress, p_w represents the pressure in the pores of the rock, α represents the coefficient proposed by Biot, and δ represents the constant proposed by Kronecker.

2.4. XFEM Approximation

Belytschko and Black [43] introduced the XFEM to overcome the shortcomings of the FEM when simulating the propagation of various discontinuities.

In order to characterize a fracture, it is necessary to construct discontinuous displacement functions. A discontinuous displacement function consists of an asymptotic function and a discontinuity function [28]. The displacement vector u is written as follows:

$$u = \sum_{I=1}^{N} N_I(x)\left[u_I + H(x)a_I + \sum_{\alpha=1}^{4} F_\alpha(x) b_I^\alpha\right] \tag{4}$$

where u_I represents the continuous node displacement vector associated with the element in the conventional finite element solution, I represents the total number of nodes within the element, $N_I(x)$ represents the conventional node shape function, $H(x)$ represents the correlation function of the discontinuous jumps of the fracture propagation element, a_I represents the additional degrees of freedom of the fracture propagation element, $F_\alpha(x)$ represents the additional function of the fracture tip, and b_I^α represents the additional degrees of freedom of the fracture tip element.

The values of the discontinuity jump functions are as follows [38]:

$$H(x)\begin{cases} 1 & (x - x^*) \cdot \mathbf{n} \geq 0 \\ -1 & \text{otherwise} \end{cases} \tag{5}$$

where x represents the node under investigation, x^* represents the node on the fracture with the shortest distance from x, and $\mathbf{n}$ represents the unit outward normal vector at x^*.

The additional function $F_\alpha(x)$ for the tip of the fracture (Figure 2) is given by the following equation [38]:

$$F_\alpha(x) = [\sqrt{r}\sin\frac{\theta}{2}, \sqrt{r}\cos\frac{\theta}{2}, \sqrt{r}\sin\theta\sin\frac{\theta}{2}, \sqrt{r}\sin\theta\cos\frac{\theta}{2}] \tag{6}$$

where r and θ represent the values of the polar coordinates in the localized frame of the element at the tip of the fracture.

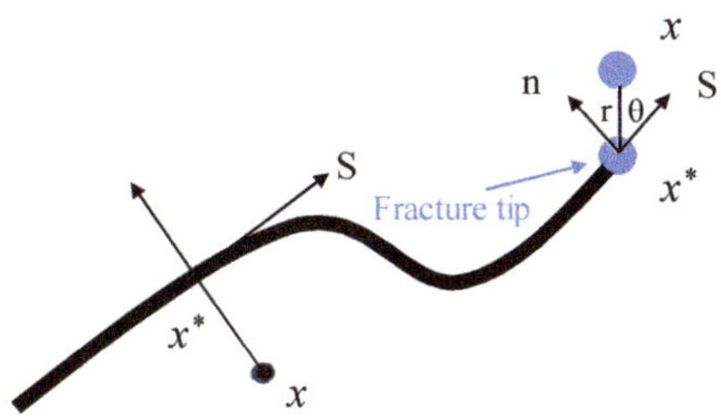

Figure 2. Schematic diagrams of smooth fractures and their normal and tangential coordinates for elements at different locations.

To accurately describe the singularity of a fracture tip through modeling, it is essential to constantly track the location of the fracture propagation. As shown in Figure 3, the discontinuities of the fracture elements can be characterized in terms of initial real nodes and phantom nodes covering the initial real nodes. Both types of nodes are independent. Therefore, when an element is penetrated by a fracture, it is divided into two parts consisting of two types of nodes, while the phantom nodes are constrained by the initial real nodes in the complete element. In order to construct a complete set of interpolations, certain fracture elements belonging to the real domain Ω_0 will be extended into the phantom domain Ω_p. Displacements of Ω_0 can be interpolated through the degree of freedom of Ω_p. The jump in the displacement field is realized by integrating from the real node side to the fracture region [28,38,40].

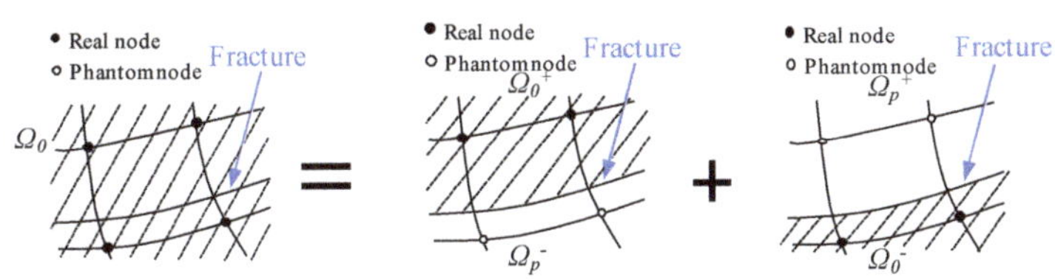

Figure 3. Schematic for introducing virtual nodes superimposed on the initial nodes to characterize the discontinuities of fracture elements.

2.5. The Mechanism of Fracture Failure

The mechanism of fracture failure consists of two processes: (1) the initiation of a fracture and (2) the evolution of fracture damage.

In elastic mechanics, failure initiation is defined as a weakening of the stiffness of a material.

During hydraulic fracturing, the maximum stress criterion is usually adopted to determine whether the stress in the reservoir rock meets the fracture initiation conditions, that is, i.e., the tensile strength of the rock is reached. Therefore, the initiation criterion can be derived from the following equation [28]:

$$f = \left\{ \frac{\langle \sigma_{\max} \rangle}{\sigma^0_{\max}} \right\} \tag{7}$$

where f represents the ratio of the maximum stress value, $\sigma^0_{\max}$ represents the allowable maximum stress value, and <> represents Macaulay Brackets. When f reaches 1, damage is initiated.

During the propagation of hydraulic fracturing, we use the Bengeggagh–Kenane law (BK law) as a criterion for describing the evolution of fracture damage [26,27]. The criterion proposed by Bengeggagh–Kenane [27,28] can be written as follows:

$$G_{equivC} = G_{IC} + (G_{IIC} - G_{IC})\left(\frac{G_{IIC} + G_{IIIC}}{G_{IC} + G_{IIC} + G_{IIIC}}\right) \tag{8}$$

where G_{equivC} represents the critical energy release rate of the fracture, G_{IC} represents the energy release rate for the tensile failure of the fracture, G_{IIC} represents the energy release rate for the shear failure of the fracture due to sliding, and G_{IIIC} is the energy release rate for the shear failure of the fracture due to tearing. The fracture propagation demand is equal to the energy release rate of the fracture between the fracture tip node and the critical value.

3. Model Establishment and Verification

3.1. Establishment of a Multiple-Fracture Development Model and the Parameterization of Different Cases

In order to explore the influences of geological and operational factors on fracture morphology and flow rate distribution during multiple-fracture development, a 2D multiple-fracture development model based on the XFEM was established, and five fractures were embedded in a reservoir. Horizontal wells were oriented in the direction of the horizontal stress with the minimum value, and five perforation clusters were oriented in the direction of the horizontal stress with the maximum value. The model was a square with a side length of 140 m (Figure 4). The number of grid elements in the model was 19,600. Table 1 establishes the basic properties of the reservoir. The conditions in the model with respect to the displacement boundary are shown in Figure 4. The displacement around the boundary was constrained to zero in the vertical direction. Incompressible fluid was injected at a constant rate into the five perforation clusters.

Table 1. Constant parameters of the model used in the simulation.

	Unit	Value
Maximum horizontal stress	MPa	98
Minimum horizontal stress	MPa	90
Poisson's modulus	-	0.14
Pore pressure	MPa	38
Young's modulus	GPa	23
Reservoir permeability	mD	1.4
Reservoir porosity	-	0.071
Tensile strength	MPa	12

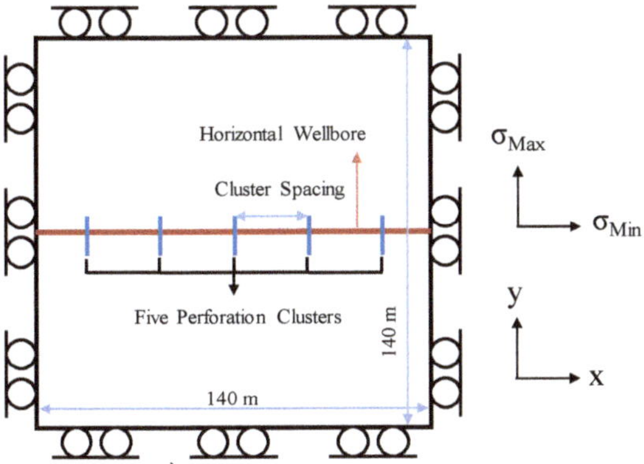

Figure 4. Simulation model of a single horizontal well with five perforation clusters and the corresponding boundary conditions.

Using the developed model, it was possible to explore the simultaneous initiation and development of five fractures in a single stage. Different geological and operational factors were considered in this study, and the parameter settings for different cases are shown in Table 2.

Table 2. Design parameters for nine cases.

Case	Cluster Spacing (m)	In Situ Stress Difference (MPa)	Fracture Fluid Viscosity (mPa·s)	Injection Rate (m^3/min)
Case 1	8	0	40	14
Case 2	8	8	40	14
Case 3	8	16	40	14
Case 4	9	8	40	14
Case 5	10	8	40	14
Case 6	8	8	1	14
Case 7	8	8	300	14
Case 8	8	8	40	12
Case 9	8	8	40	16

3.2. The Validation of Fracture Propagation

The performance of the fracture propagation model based on the XFEM was evaluated by using experimental results. Jiang et al. [44] conducted triaxial experiments to investigate the effect of oriented perforation on fracture development. Hydraulic fracturing was performed on cubic concrete samples measuring 300 mm × 300 mm × 300 mm. A borehole was drilled in the center of each sample, with a 60° angle between the borehole and the maximum horizontal principal stress. The results showed that when the direction of perforation did not coincide with the maximum horizontal principal stress, the fracture was deflected. To verify the validation of our model, the fracture propagation results simulated with the experimental parameters (Table 3) were compared with the results of the experiment (Figure 5). The results of the model consisted of the direction of fracture propagation, the morphology, and element displacement values (U). The element displacement values were able to represent the degree of fracture opening at different locations. The results of the numerical simulation showed that in the square model with a side length of 0.3 m, the hydraulic fracture first initiated along the 60° perforation azimuth with a larger displacement vector value at the location of the initiation, and then the fracture was deflected, was shifted to the direction of the maximum principal stress, and propagated to the model boundary. Fractures in the vicinity of the initiation site had large displacement values that gradually decreased after the fracture had turned. The numerical results were in general agreement with the fracture deflection and morphology observed in the experimental results of Jiang et al. [44].

Table 3. Basic parameters used for model verification.

	Unit	Value
Maximum horizontal stress	MPa	6
Minimum horizontal stress	MPa	1
Poisson's modulus	-	0.23
Young's modulus	GPa	8.402
Reservoir permeability	mD	0.1
Reservoir porosity	%	1.85
Tensile strength	MPa	2.59
Injection rate	m^3/s	2.1×10^{-9}
Fracturing fluid viscosity	mPa·s	73
The orientation between perforation and the maximum horizontal principal stress	°	60

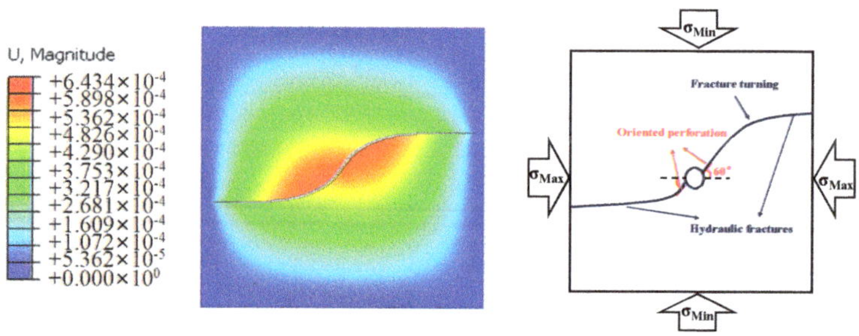

(**a**) (**b**)

Figure 5. Comparison between the numerical model and the experimental result (unit: m). (**a**) the numerical simulation result, (**b**) the schematic diagram of experimental results.

4. Results and Discussion

In this section, the developed model was utilized to investigate the simultaneous initiation and development of five fractures in a single stage. When a certain amount of injected fluid flowed into the five fractures, the amount of fluid flowing into each fracture was redistributed and constantly changed. By comparing the morphology of the fractures and the flow rate distribution, the influence of stress-shadow effects during the development of multiple fractures can be discussed. Here, different geological and operational factors are considered, as shown in Table 3.

4.1. The Effect of In Situ Stress Difference (Cases 1–3)

Differences in in situ stresses are an important factor in fracture propagation. The direction of fracture propagation is usually the direction of the horizontal stress with the minimum value. However, under the condition of the simultaneous development of multiple fractures, the stress field around the fracture tip region can be disturbed due to stress-shadow effects. Then, the fracture undergoes different degrees of deflection. In order to explore the effects of in situ stress differences on the development of multiple fractures, three values of stress difference were used, i.e., 0 MPa, 8 MPa, and 16 MPa.

The morphology of the five fractures after 240 s of fluid injection is shown in Figure 6. Figure 6 shows that brighter-colored locations represent fractures with higher displacement values. It was observed that the fractures (except for the center fracture) developed symmetrically and had the same morphology. Regarding the propagation process of the fractures, the first five fractures extended along the direction of the horizontal stress with the maximum value. Then, based on the stress-shadow effects, the side fracture and sub-center fracture deviated from the center fracture. Apparently, the center fracture was compressed by the sub-center fractures, and little deflection occurred. In addition, the deflections of the side and sub-center fractures decreased with the increase in stress difference. In particular, Figure 6a shows that at a stress difference of 0 MPa, all fractures except the center fracture underwent significant deflection. For the sub-center fractures, the bottom tip was connected to the side fracture. The presence of this phenomenon greatly reduced the effectiveness of the side and sub-center fractures.

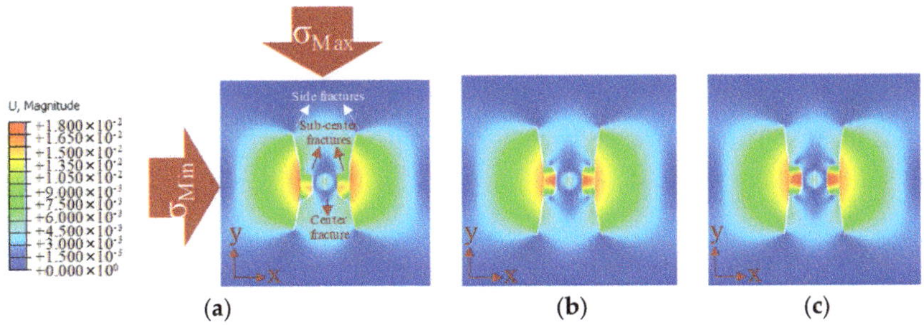

Figure 6. Displacement values and fracture geometries for three different in situ stress differences (unit: m). (**a**) The value of the in situ stress difference was 0 MPa; (**b**) the value of the in situ stress difference was 8 MPa; (**c**) the value of the in situ stress difference was 16 MPa.

Figure 7 depicts the length of each fracture after 240 s of fluid injection. It can be seen that the length of the sub-center fracture significantly increased as the in situ stress difference increased. The length variations in the other fractures were negligible. This was because the magnitude of deformation in the side fractures was suppressed with higher stress differences. Also, less energy was required for fracture propagation along the direction of the horizontal stress with the maximum value compared to that in the other directions. The length of a reasonably oriented fracture was larger for the same fluid flow rate. When the deflection amplitude of the side fractures decreased, the stress field around the side fractures enhanced the effect of stress interference on the sub-center fractures and central fracture. As a result, the length of the sub-center fracture did not significantly increase as the deflection amplitude decreased. On the contrary, the length of the otherwise flat center fracture decreased.

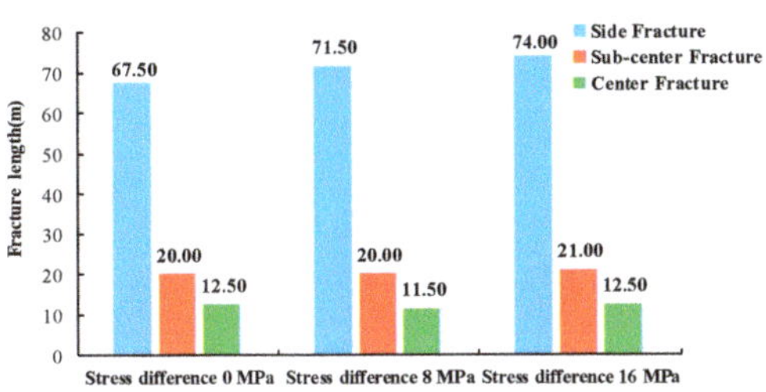

Figure 7. Fracture lengths for simultaneous development with three different in situ stress differences.

Figure 8 shows the amount of fluid injected into each fracture. It can be noticed that five fractures were initiated at the beginning of the fluid injection, and the fluid flow rate inside each fracture rapidly increased. When the fractures were in the propagation stage, the fluid flow rate in each fracture tended to stabilize. In contrast to the side fractures, the development of the sub-center fractures and center fracture was limited by stress-shadow effects and accounted for a significantly smaller proportion of the total fluid volume.

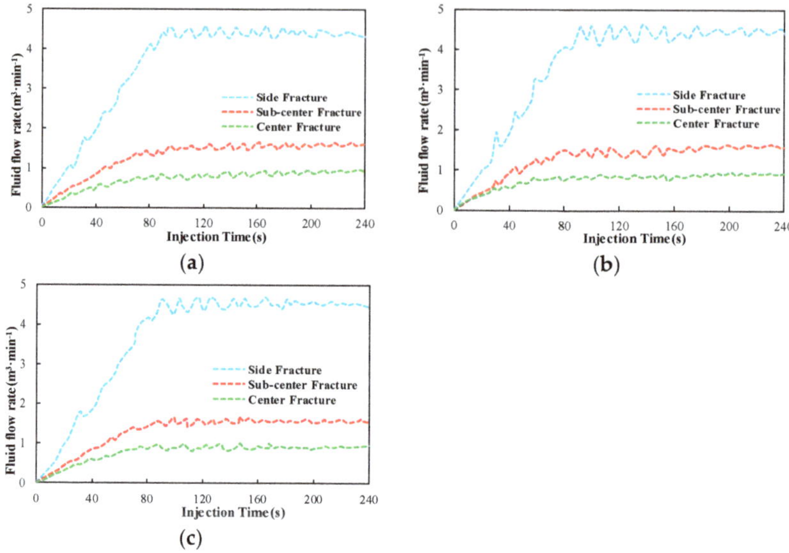

Figure 8. Fluid flow rates of simultaneously propagating fractures with three different in situ stress differences. (**a**) The value of the in situ stress difference was 0 MPa; (**b**) the value of the in situ stress difference was 8 MPa; (**c**) the value of the in situ stress difference was 16 MPa.

For example, with a stress difference of 0 MPa, the proportions of fluid entering the center fracture, the single sub-center fracture, and the single side fracture were 7.22%, 12.63%, and 33.76%, respectively, after 240 s of fluid injection (Figure 9). The side fractures were hardly affected by stress-shadow effects and could develop well. However, the center fracture was most affected by stress-shadow effects, and less fluid entered it.

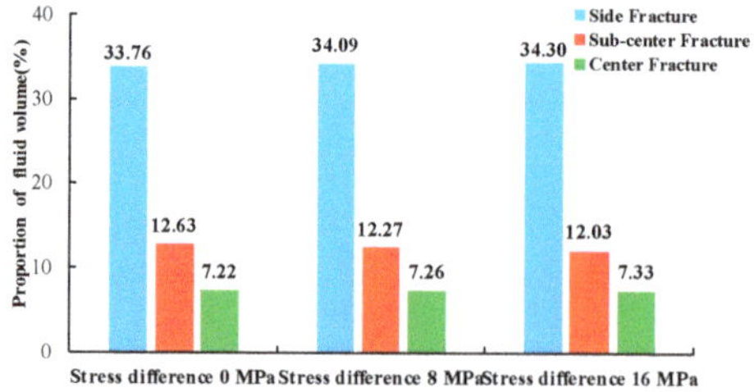

Figure 9. Proportions of fluid volume entering the fractures after 240 s of fluid injection for three different in situ stress difference conditions.

In general, as the in situ stress difference increased, the fractures deflected by the stress-shadow effect gradually returned to the direction of the horizontal stress with the maximum value. The propagation of the internal fractures was more strongly inhibited. However, there was little change in the fluid within each fracture.

4.2. The Effects of Cluster Spacing (Cases 2, 4, and 5)

In actual hydraulic fracturing processes, the geological parameters are usually fixed. It is more practical to reduce the effects of stress interferences by optimizing the operating parameters. Cluster spacing plays an important role in the uniform development of multiple fractures. In order to discuss the specific effects of cluster spacing on the morphology and fluid flow of each fracture, three different cluster spacings (8 m, 9 m, and 10 m) were set up in the model.

Figure 10 shows the morphology of the five fractures after 240 s of fluid injection with three cluster spacings. In Figure 10, brighter-colored locations represent fractures with higher displacement values. As the spacing of the perforation clusters increased, the degree of deflection of the side fractures increased. When the clusters were spaced 10 m apart, the side fractures propagated in a direction almost away from the center fracture at the location of initiation. The sub-center fracture initially propagated in the direction of the horizontal stress with the maximum value. Then, the direction was away from the center fracture. The center fracture propagated in a straight line along the direction of the horizontal stress with the maximum value. The development of the center fracture significantly increased with the increase in the cluster spacing.

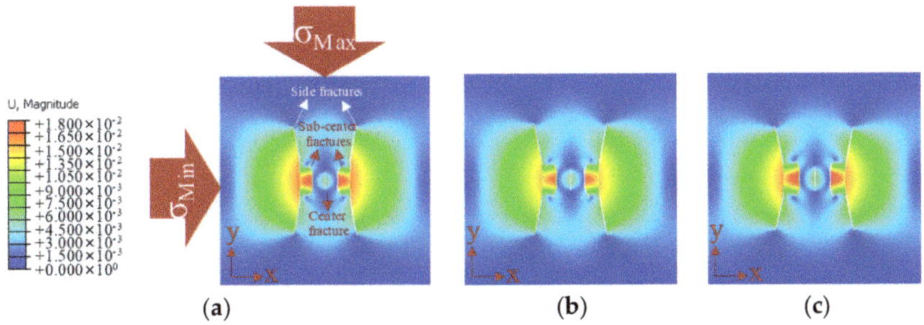

Figure 10. Displacement values and fracture geometries for three different cluster spacings (unit: m). (**a**) The spacing of the perforation clusters was 8 m; (**b**) the spacing of the perforation clusters was 9 m; (**c**) the spacing of the perforation clusters was 10 m.

Figure 11 depicts the length of each fracture after 240 s of fluid injection. The length of the side fractures increased and then decreased as the spacing of the perforation cluster increased. The length of the sub-center fracture first increased and then remained stable. The length of the center fracture continuously increased. It was observed that the larger cluster spacing contributed to the uniform development of multiple fractures, which was favorable for improving the fracturing effect.

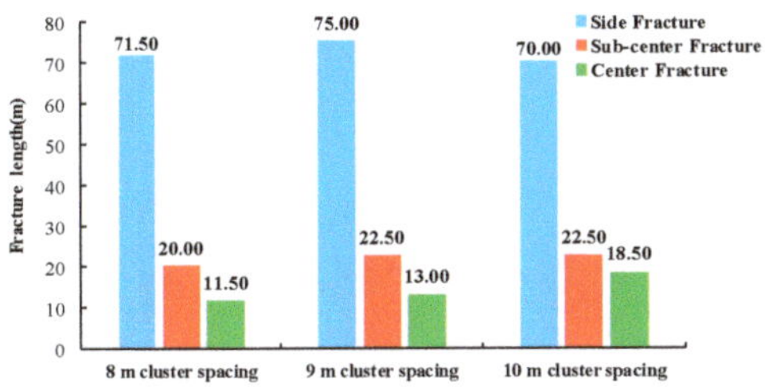

Figure 11. Fracture length of simultaneously developed fractures with three different cluster spacings.

Figure 12 shows the amount of fluid injected into each fracture. The proportion of the fluid volume in each fracture after 240 s of fluid injection is shown in Figure 13. By comparing the change curves of the fluid flowing into each fracture, it was found that the proportion of each type of fracture gradually stabilized after reaching the maximum value. Combined with the fracture morphology after 240 s of fluid injection (Figure 10), it was found that the initial propagation of the side fractures inhibited the development of the sub-center fractures and center fracture. However, the stress-shadow effects of the sub-center fractures and center fracture were weakened with the increase in the cluster spacing and side fracture deflection. The fracture development was improved to different degrees. After 240 s of fluid injection, the proportion of fluid in the side, sub-center, and center fracture changed from 34.16%, 12.33%, and 7.02% to 30.9%, 14.28%, and 9.64%, respectively, with the increase in the cluster spacing. The uniformity of the flow distribution in each fracture was significantly improved.

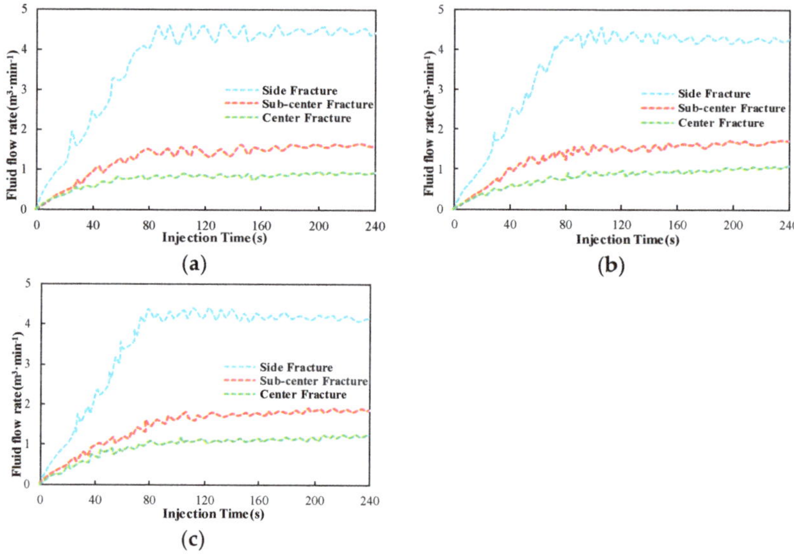

Figure 12. Fluid flow rates of simultaneously propagating fractures with three different cluster spacings. (**a**) The spacing of the perforation clusters was 8 m; (**b**) the spacing of the perforation clusters was 9 m; (**c**) the spacing of the perforation clusters was 10 m.

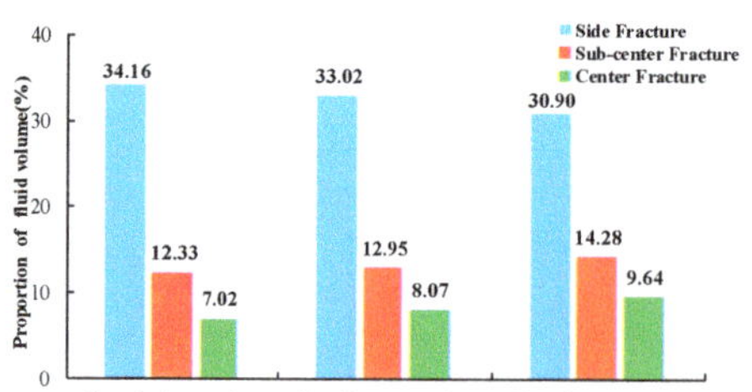

Figure 13. Proportions of fluid volume entering the fractures after 240 s of fluid injection for three different cluster spacings.

In conclusion, increasing the spacing of the perforation clusters attenuated stress interferences and improved the uniformity of the flow rate distribution and fracture development.

4.3. The Effects of Fracturing Fluid Viscosity (Cases 2, 6, and 7)

Fracturing fluid viscosity is an important operational parameter that has a significant impact on the fracturing pattern. In order to discuss the various effects of the fracturing fluid viscosity on the fracture patterns and fluid flow in each fracture, three different values of fracturing fluid viscosity (1 mPa·s, 40 mPa·s, and 300 mPa·s) were set for discussion.

Figure 14 shows the morphology of the five fractures after 240 s of fluid injection. In Figure 14, brighter-colored locations represent fractures with higher displacement values. From the fracture development morphology, it can be seen that the degree of deflection of the side fractures increased as the viscosity changed from 1 mPa·s to 40 mPa·s. The morphology of the sub-center fractures and the center fracture remained almost unchanged. However, the morphology of each fracture changed significantly as the viscosity changed from 40 mPa·s to 300 mPa·s. The side and sub-center fractures propagated almost completely in the direction of the horizontal stress with the maximum value, and the length of the center fracture decreased significantly.

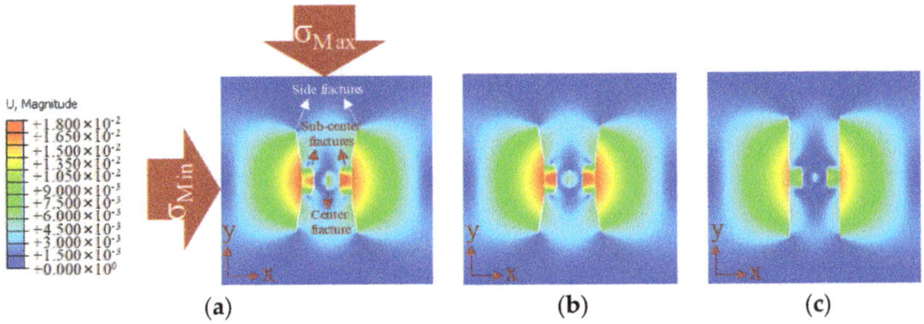

Figure 14. Displacement values and fracture geometries for three different fracturing fluid viscosities (unit: m). (**a**) The value of the fracturing fluid viscosity was 1 mPa·s; (**b**) the value of the fracturing fluid viscosity was 40 mPa·s; (**c**) the value of the fracturing fluid viscosity was 300 mPa·s.

Figure 15 depicts the length of each fracture after 240 s of fluid injection. The length of each fracture did not change much as the viscosity changed from 1 mPa·s to 40 mPa·s. The deflection angle of the side cracks increased with increasing viscosity. The area of the fracture network also increased. When the viscosity was changed from 40 mPa·s to 300 mPa·s, the length of the side fractures significantly increased from 71.5 m to 80 m. On the contrary, the lengths of the sub-center fractures and the center fracture decreased by 5 m and 2.5 m, respectively. The results showed that the side fractures propagated along the direction of the horizontal stress with the maximum value when the viscosity was very high. The propagation process was easier and required less energy. Meanwhile, the development of the sub-center fractures and center fracture was inhibited, which was influenced by the larger stress-shadow effect. Therefore, this propagation was not conducive to the uniform development of multiple fractures.

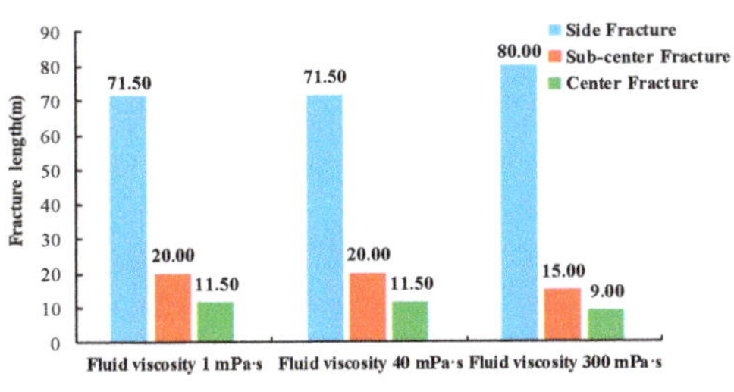

Figure 15. Fracture lengths of simultaneously developed fractures with three different fracturing fluid viscosities.

Figure 16 shows the fluid flow rate into each fracture during fluid injection. The proportions of fluid flow into each fracture after 240 s of fluid injection are shown in Figure 17. It can be seen that when the viscosity was 1 mPa·s, the fluid flow rate into the sub-center fracture was significantly larger than that into the center fracture. With the increase in viscosity, the fluid flow into the sub-center fracture tended to be the same as that in the center fracture. When the viscosity increased, the fluid volume proportion in the side fracture increased from 32.35% to 39.85%, while the fluid volume proportion in the sub-center and center fractures decreased by 5.87% and 3.26%, respectively (Figure 17).

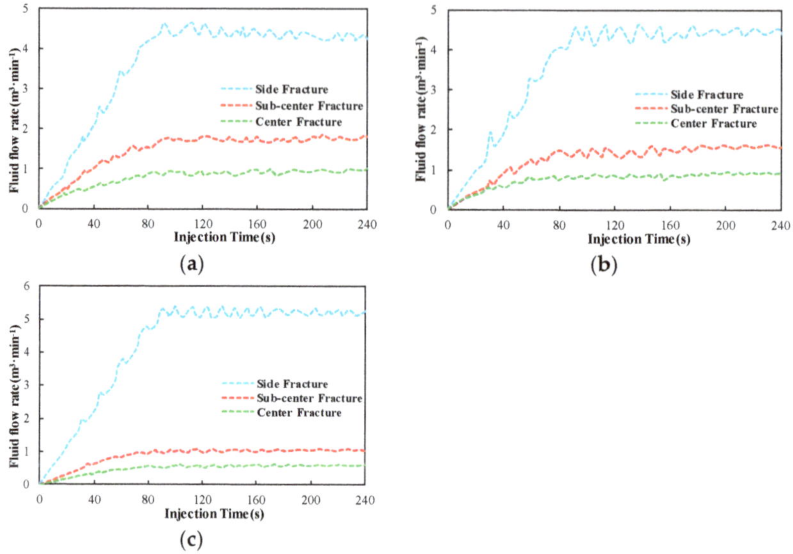

Figure 16. Fluid flow rates of simultaneously propagating fractures with three different fracturing fluid viscosities. (**a**) The value of the fracturing fluid viscosity was 1 mPa·s; (**b**) the value of the fracturing fluid viscosity was 40 mPa·s; (**c**) the value of the fracturing fluid viscosity was 300 mPa·s.

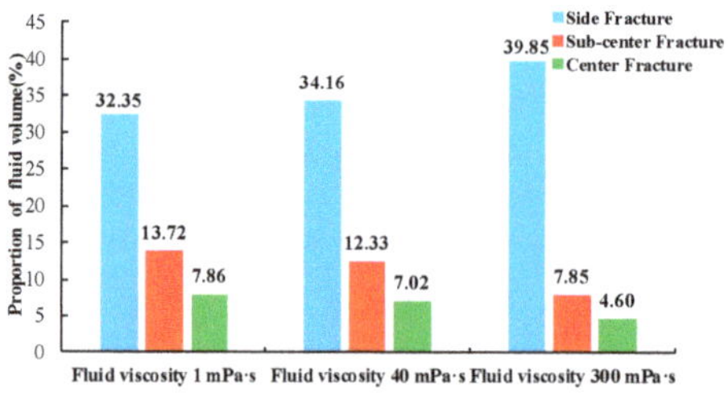

Figure 17. Proportions of fluid volume entering the fractures after 240 s of fluid injection for three different fracturing fluid viscosities.

In summary, the stress-shadow effect increased with increasing viscosity. However, if the viscosity is increased appropriately, a larger fracturing network can be formed.

4.4. The Effect of Injection Rate (Cases 2, 8, and 9)

The injection rate is another important operating parameter. In order to discuss the various effects of the injection rate on the fracture morphology and fluid flow in each fracture, three different injection rates (12 m^3/min, 14 m^3/min, and 16 m^3/min) were set for discussion.

Figure 18 shows the morphology of the five fractures after 240 s of fluid injection at three injection rates. In Figure 18, brighter-colored locations represent fractures with higher displacement values. When the injection rate was increased, the length of the side fractures increased significantly, while the deflection of the side fractures did not change much. The length and deflection angle of the sub-center fractures and the center fracture hardly changed. It was observed that the development of the major fractures increased with the increase in the injection rate. On the contrary, the other fractures remained largely unchanged. Unfortunately, the unevenness of fracture development increased.

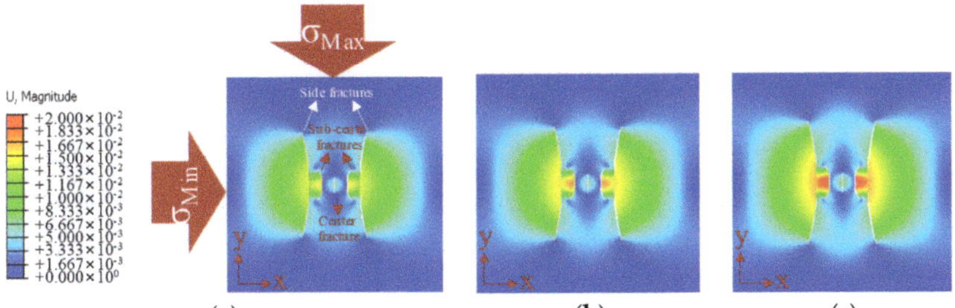

Figure 18. Displacement values and fracture geometries for three different injection rates (unit: m). (**a**) The value of the injection rate was 12 m^3/min; (**b**) the value of the injection rate was 14 m^3/min; (**c**) the value of the injection rate was 16 m^3/min.

Figure 19 depicts the length of each fracture after 240 s of fluid injection. As the injection rate increased, the length of the side fractures increased from 70 to 80 m, while the length of the other fractures remained almost constant (Figure 19).

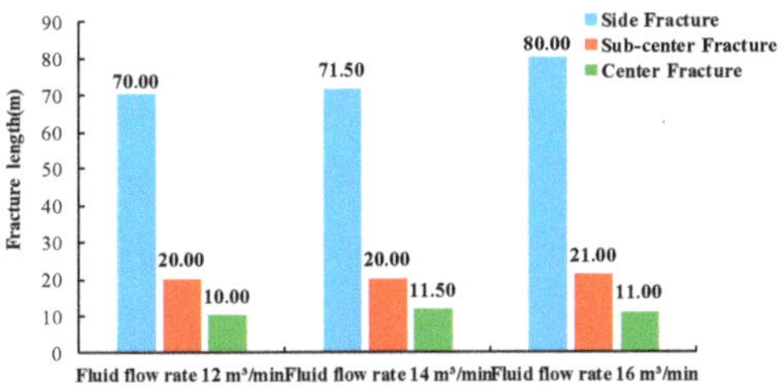

Figure 19. Fracture lengths of simultaneously developed fractures with three different injection rates.

Figure 20 shows the flow rate in each fracture during fluid injection. By comparing the change curves when using different injection rates, it was found that the volume of fluid entering each fracture tended to increase with the increase in the injection rate. The proportion of fluid volume entering each fracture after 240 s of fluid injection is shown in Figure 21. At different injection rates, the fluid volume proportion in each fracture remained almost constant.

In conclusion, from the point of view of fracture morphology, the length of the side fractures significantly increased with the increase in the injection rate. In contrast, the change in the length of the internal fractures was negligible. Although the area of the fracture network increased, the degree of inhomogeneous development between fractures increased. In addition, the proportion of fluid volume in each fracture changed little when the injection rate was increased. Therefore, the injection rate had a small effect on the stress-shadow effect, as well as on the distribution of the flow rate between fractures.

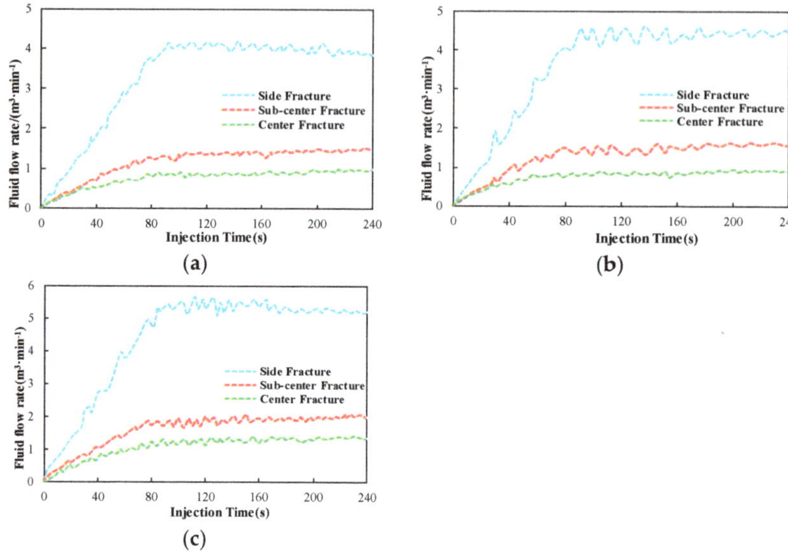

Figure 20. Fluid flow rates of simultaneously propagating fractures with three different injection rates. (**a**) The value of the injection rate was 12 m^3/min; (**b**) the value of the injection rate was 14 m^3/min; (**c**) the value of the injection rate was 16 m^3/min.

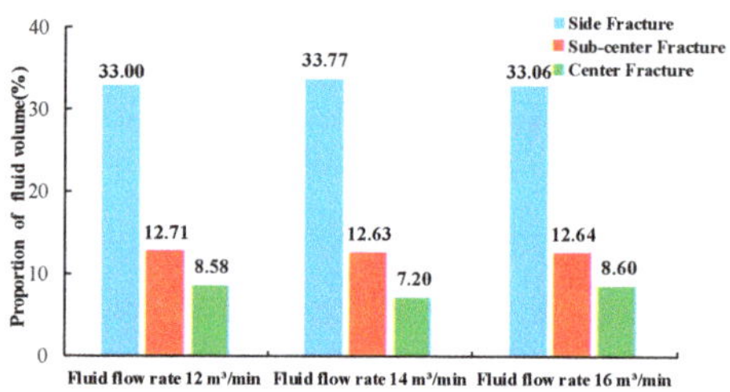

Figure 21. Proportions of fluid volume entering the fractures after 240 s of fluid injection for three different injection rates.

5. Discussion and Conclusions

In this study, a two-dimensional hydraulic fracturing propagation model was established based on the XFEM. The effects of geological and operational factors on the simultaneous development of multiple fractures were investigated. The study helps to set reasonable operation parameters for improving the uniformity of multiple fractures and forming a complex fracture network, which will be conducive to improving shale gas recovery. The following conclusions were drawn:

- The influence of the in situ stress difference on the flow rate distribution among multiple fractures is not obvious. However, with the increase in the in situ stress difference, side fractures with larger deflections gradually propagate along the direction of the horizontal stress with the maximum value. In addition, the fracture length significantly increases.
- The cluster spacing can effectively affect the stress interference between multiple fractures. The uniformity of the fracture morphology and flow rate distribution can be significantly improved by increasing the cluster spacing.
- As the viscosity of the fracturing fluid increases, the stress-shadow effect is further enhanced, and more fluid flows into the dominant fracture. When the viscosity increases

from 1 mPa·s to 40 mPa·s, the deflection angle of the side fractures increases, as does the fracture network area. When the viscosity reaches 300 mPa·s, the development of internal fractures is greatly inhibited, and the side fractures are no longer deflected.
- A higher injection rate has little effect on the inflow of fluid into the internal fractures and on their morphology. On the other hand, with the same proportion of the fluid volume, the length of the side fractures significantly increases, favoring the formation of a larger fracture network area.

It can be seen from the results that setting a suitable perforation cluster direction, increasing the cluster spacing, and using a medium (e.g., 40 mPa·s) fracturing fluid viscosity according to the in situ stress conditions can help weaken the influence of the stress-shadow effect caused by the simultaneous development of multiple fractures and promote the uniform development of fractures formed by multiple perforation clusters. Increasing the injection rate and setting the appropriate fracturing fluid viscosity will help increase the degree of development of each fracture and form a certain degree of deflection, which will be conducive to increasing the complexity and coverage area of the fracture network.

Of course, the shortcomings of this study must be pointed out. Currently, the formation of complex fracture networks through hydraulic fracturing requires attention to the longitudinal development of fractures. The height of multiple fractures in the 2D hydraulic fracturing model was assumed to be the same as the height of the reservoir. In order to know the true evolution of the height of multiple fractures, in future work, we will develop a three-dimensional hydraulic fracture propagation model; then, the morphologies of multiple fractures can be studied more comprehensively.

Author Contributions: Methodology: E.P.; Software: H.Z. and P.Z.; Data curation: Z.J.; writing—original draft preparation, review, and editing: Y.W., Q.Z. and G.F. All authors have read and agreed to the published version of the manuscript.

Funding: This study was financially supported by the National Natural Science Foundation of China (Nos. 52192622, 52374004, and 52204005) and the Natural Science Foundation of Sichuan Province, China (Grant No. 2022NSFSC0193).

Data Availability Statement: Data is contained within the article.

Conflicts of Interest: Authors Enshun Ping, Yuzhong Wang and Qingjie Zhao were employed by the company CNPC Bohai Drilling Engineering Company Limited. The remaining authors declare that the research was conducted in the absence of any commercial or financial relationships that could be construed as a potential conflict of interest.

References

1. Li, L.; Liu, J.; Wang, W.; Zhao, P.; Lu, Y. Variational Reservoir Characteristics of the Wufeng–Longmaxi Formation from Different Sedimentary Region and Its Implications in Southeastern Sichuan Basin, China. *ACS Omega* **2023**, *8*, 20684–20696. [CrossRef] [PubMed]
2. Liu, J.; Yao, Y.; Liu, D.; Pan, Z.; Cai, Y. Comparison of Three Key Marine Shale Reservoirs in the Southeastern Margin of the Sichuan Basin, SW China. *Minerals* **2017**, *7*, 179. [CrossRef]
3. Zhao, P.; Liu, J.; Elsworth, D. Numerical study on a multifracture enhanced geothermal system considering matrix permeability enhancement induced by thermal unloading. *Renew. Energ.* **2023**, *2023*, 33–44. [CrossRef]
4. Liu, J.; Bai, X.; Elsworth, D. Evolution of pore systems in low-maturity oil shales during thermal upgrading-Quantified by dynamic SEM and machine learning. *Pet. Sci.* **2024**, in press. [CrossRef]
5. Liu, J.; Xie, L.; Elsworth, D.; Gan, Q. CO_2/CH_4 Competitive Adsorption in Shale: Implications for Enhancement in Gas Production and Reduction in Carbon Emissions. *Environ. Sci. Technol.* **2019**, *53*, 9328–9336. [CrossRef] [PubMed]
6. Desroches, J.; Lecampion, B.; Ramakrishnan, H.; Brown, J.E.; Prioul, R. Benefits of Controlled Hydraulic Fracture Placement: Theory and Field Experiment. In Proceedings of the SPE/CSUR Unconventional Resources Conference, Calgary, AB, Canada, 30 September–2 October 2014.
7. Zhao, J.; Wang, L.; Liu, S.; Kang, Z.; Yang, D.; Zhao, Y. Numerical simulation and thermo-hydro-mechanical coupling model of in situ mining of low-mature organic-rich shale by convection heating. *Adv. Geo-Energy Res.* **2022**, *6*, 502–514. [CrossRef]
8. Wu, K. Simultaneous Multi-Frac Treatments: Fully Coupled Fluid Flow and Fracture Mechanics for Horizontal Wells. In Proceedings of the SPE Annual Technical Conference and Exhibition, New Orleans, LA, USA, 30 September–2 October 2013.

9. Wu, K.; Olson, J.E.; Balhoff, M.T.; Yu, W. Numerical Analysis for Promoting Uniform Development of Simultaneous Multiple-Fracture Propagation in Horizontal Wells. *SPE Prod. Oper.* **2016**, *32*, 41–50.
10. Warpinski, N.R.; Teufel, L.W. Influence of Geologic Discontinuities on Hydraulic Fracture Propagation. *J. Pet. Technol.* **1987**, *39*, 209–220. [CrossRef]
11. Zeng, F.; Yang, B.; Guo, J.; Chen, Z.; Xiang, J. Experimental and Modeling Investigation of Fracture Initiationfrom Open-Hole Horizontal Wells in Permeable Formations. *Rock Mech.* **2019**, *52*, 1133–1148. [CrossRef]
12. Ha, S.J.; Choo, J.; Yun, T.S. Liquid CO_2 Fracturing: Effect of Fluid Permeation on the Breakdown Pressure and Cracking Behavior. *Rock Mech.* **2018**, *51*, 3407–3420. [CrossRef]
13. Zhang, W.; Shi, X.; Jiang, S.; Cao, Q.; Wang, F.; Wang, Z.; Ge, Y.; Du, Y. Experimental study of hydraulic fracture initiation and propagation in highly saturated methane-hydrate-bearing sands. *J. Nat. Gas Sci. Eng.* **2020**, *79*, 103338. [CrossRef]
14. Liu, X.; Qu, Z.; Guo, T.; Sun, Y.; Wang, Z.; Bakhshi, E. Numerical simulation of non-planar fracture propagation in multi-cluster fracturing with natural fractures based on Lattice methods. *Eng. Fract. Mech.* **2019**, *220*, 106625. [CrossRef]
15. Qu, Z.; Tian, Y.; Li, J.; Guo, T. Numerical simulation study on fracture extension and morphology of multi-cluster staged fracturing for horizontal wells. *J. China Univ. Pet. Ed. Nat. Sci.* **2017**, *41*, 102–109.
16. Fallahzadeh, S.H.; Rasouli, V.; Sarmadivaleh, M. An Investigation of Hydraulic Fracturing Initiation and Near-Wellbore Propagation from Perforated Boreholes in Tight Formations. *Rock Mech.* **2015**, *48*, 573–584. [CrossRef]
17. Wu, K.; Olson, J.E. Mechanisms of Simultaneous Hydraulic Fracture Propagation From Multiple Perforation Clusters in Horizontal Wells. *SPE J.* **2016**, *21*, 1000–1008. [CrossRef]
18. Bunger, A.; Jeffrey, R.G.; Zhang, X. Constraints on simultaneous growth of hydraulic fractures from multiple perforation clusters in horizontal wells. *SPE J.* **2014**, *19*, 608–620. [CrossRef]
19. Wang, W.; Da, Z.; Sheng, G.; Qi, Z.; Su, Y. A review of stimulated reservoir volume characterization for multiple fractured horizontal well in unconventional reservoirs. *Adv. Geo-Energy Res.* **2017**, *1*, 54–63. [CrossRef]
20. Miller, C.; Waters, G.; Rylander, E. Evaluation of Production Log Data from Horizontal Wells Drilled in Organic Shales. In Proceedings of the SPE North American Unconventional Gas Conference and Exhibition, Woodlands, TX, USA, 14–16 June 2011.
21. Van Ketterij, R.B.; De Pater, C.J. Impact of Perforations on Hydraulic Fracture Tortuosity. *SPE Prod. Fac.* **1999**, *14*, 117–130. [CrossRef]
22. Hui, G.; Chen, Z.; Chen, S.; Gu, F. Hydraulic fracturing-induced seismicity characterization through coupled modeling of stress and fracture-fault systems. *Adv. Geo-Energy Res.* **2022**, *6*, 269–270. [CrossRef]
23. Liu, J.; Xie, L.; He, B.; Gan, Q.; Zhao, P. Influence of anisotropic and heterogeneous permeability coupled with in-situ stress on CO_2 sequestration with simultaneous enhanced gas recovery in shale: Quantitative modeling and case study. *Int. J. Greenhouse Gas Control* **2021**, *104*, 103208. [CrossRef]
24. Zhao, P.; Liu, J.; Zhang, Y. Experimental and numerical investigations on buffer performance of geofoam subjected by the impact of falling rocks with respect to different shapes. *Geotext. Geomembr.* **2023**, *51*, 108–124. [CrossRef]
25. Abdelaziz, A.; Ha, J.; Li, M.; Magsipoc, E.; Sun, L.; Grasselli, G. Understanding hydraulic fracture mechanisms: From the laboratory to numerical modelling. *Adv. Geo-Energy Res.* **2023**, *7*, 66–68. [CrossRef]
26. Guo, J.; Lu, Q.; Zhu, H.; Wang, Y.; Ma, L. Perforating cluster space optimization method of horizontal well multi-stage fracturing in extremely thick unconventional gas reservoir. *J. Nat. Gas Sci. Eng.* **2015**, *26*, 1648–1662. [CrossRef]
27. Shin, D.H.; Sharma, M.M. Factors Controlling the Simultaneous Propagation of Multiple Competing Fractures in a Horizontal Well. In Proceedings of the SPE Hydraulic Fracturing Technology Conference, Woodlands, TX, USA, 4–6 February 2014.
28. Wang, B.; Zhou, F.; Wang, D.; Liang, T. Numerical simulation on near-wellbore temporary plugging and diverting during refracturing using XFEM-Based CZM. *J. Nat. Gas Sci. Eng.* **2018**, *55*, 368–381. [CrossRef]
29. Liu, Y.; Tang, D.; Xu, H.; Zhao, T.; Hou, W. Effect of interlayer mechanical properties on initiation and propagation of hydraulic fracturing in laminated coal reservoirs. *J. Nat. Gas Sci. Eng.* **2022**, *208*, 109381. [CrossRef]
30. Huang, C.; Zhu, H.; Wang, J.; Han, J.; Zhou, G.; Tang, X. A FEM-DFN model for the interaction and propagation of multi-cluster fractures during variable fluid-viscosity injection in layered shale oil reservoir. *Pet. Sci.* **2022**, *19*, 2796–2809. [CrossRef]
31. Zhu, H.; Xu, X.; Zhong, A.; Zhang, Q. Numerical simulation of evenly propagating hydraulic fractures with smaller cluster spacing in the horizontal well YYP1 for deep shale oil in the Shengli Oilfield. *Oil Gas Geology.* **2022**, *43*, 229–240.
32. Guo, J.; Zhao, X.; Zhu, H.; Zhang, X.; Pan, R. Numerical simulation of interaction of hydraulic fracture and natural fracture based on the cohesive zone finite element method. *J. Nat. Gas Sci. Eng.* **2015**, *25*, 180–188. [CrossRef]
33. Wang, Y.; Javadi, A.; Fidelibus, C. A hydro-mechanically-coupled XFEM model for the injection-induced evolution of multiple fractures. *Int. J. Numer. Anal. Methods Geomech.* **2023**, *47*, 1539–1558. [CrossRef]
34. Dahi-Taleghani, A.; Olson, J.E. Numerical modeling of multistranded-hydraulic-fracture propagation: Accounting for the interaction between induced and natural fractures. *SPE J.* **2011**, *16*, 575–581. [CrossRef]
35. Gordeliy, E.; Peirce, A. Implicit level set schemes for modeling hydraulic fractures using the XFEM. *Comput. Method Appl. Mech.* **2013**, *266*, 125–143. [CrossRef]
36. Lecampion, B. An extended finite element method for hydraulic fracture problems. *Commun. Numer. Methods Eng.* **2009**, *25*, 121–133. [CrossRef]
37. Leonhart, D.; Meschke, D. Extended finite element method for hydro-mechanical analysis of crack propagation in porous materials. *Proc. Appl. Math. Mech.* **2011**, *11*, 161–162. [CrossRef]

38. Wang, H. Numerical investigation of fracture spacing and sequencing effects on multiple hydraulic fracture interference and coalescence in brittle and ductile reservoir rocks. *Eng. Fract. Mech.* **2016**, *157*, 107–124. [CrossRef]
39. Ma, X.; Zeng, Q. Influence of multi-perforation fracturing on the regularity of hydraulic fractures propagation. *Nat. Gas Geosci.* **2022**, *33*, 312–323.
40. Gordeliy, E.; Peirce, A. Coupling schemes for modeling hydraulic fracture propagation using the XFEM. *Comput. Method Appl. Mech.* **2013**, *253*, 305–322. [CrossRef]
41. Cruz, F.; Rpehl, D.; Vargas, E.D. An XFEM implementation in Abaqus to model intersections between fractures in porous rocks. *Comput. Geotech.* **2019**, *112*, 135–146. [CrossRef]
42. Yang, Q.; Li, B.; Shao, M.; Zhang, H. Production performance of perforation clusters during multistage fracturing in shale gas reservoirs. *ACS Omega.* **2021**, *6*, 26231–26238. [CrossRef]
43. Belytschko, T.; Black, T. Elastic crack growth in finite elements with minimal remeshing. *Int. J. Numer. Meth. Eng.* **1999**, *45*, 601–620. [CrossRef]
44. Jiang, H.; Chen, M.; Zhang, G.; Jin, Y.; Zhang, Z.; Zhu, G. Impact of oriented perforation on hydraulic fracture initiation and propagation. *Chin. J. Rock Mech. Eng.* **2009**, *7*, 1321–1326.

energies

MDPI

Article

Characterization of the Macroscopic Impact of Diverse Microscale Transport Mechanisms of Gas in Micro-Nano Pores and Fractures

Yintao Dong [1,*], Laiming Song [1], Fengpeng Lai [2], Qianhui Zhao [2], Chuan Lu [1], Guanzhong Chen [1], Qinwan Chong [1], Shuo Yang [1] and Junjie Wang [1]

1 CNOOC Research Institute, Beijing 100028, China
2 School of Energy Resources, China University of Geosciences, Beijing 100083, China
* Correspondence: dongyt4@coori.com.cn; Tel.: +86-010-89913636

Abstract: The objective of this study is to construct a refined microscopic transport model that elucidates the transport mechanisms of gas flow within micro-nano pores and fractures. The collective impact of various microscopic transport mechanisms was explained through the apparent permeability model, specifically related to gases such as methane and carbon dioxide, within the shale matrix. The apparent permeability models, taking into account microscopic transport mechanisms such as slippage flow, Knudsen diffusion, transition flow, and surface diffusion, were established individually. Subsequently, the influencing factors on apparent permeability were analyzed. The results demonstrate that the apparent permeability of the shale reservoir matrix is significantly influenced by pore pressure, temperature, pore size, and total organic carbon (TOC). As pressure decreases, the apparent permeability of Knudsen diffusion and surface diffusion increases, while the apparent permeability of slippage flow decreases. In addition, the apparent permeability of the reservoir matrix initially decreases and then increases. With increasing temperature, the apparent permeability of slippage flow, Knudsen diffusion, and surface diffusion all increase, as does the apparent permeability of the reservoir matrix. As pore size increases, the apparent permeability of surface diffusion and Knudsen diffusion decreases, while the apparent permeability of slippage flow and the reservoir matrix increases. Furthermore, an increase in TOC leads to no change in the apparent permeability of slippage flow and Knudsen diffusion, but an increase in the apparent permeability of surface diffusion and the reservoir matrix. The model presented in this paper enhances the multi-scale characterization of gas microflow mechanisms in shale and establishes the macroscopic application of these micro-mechanisms. Moreover, this study provides a theoretical foundation for the implementation of carbon capture, utilization, and storage (CCUS) in shale gas production.

Keywords: micro-nano pore; shale gas; carbon dioxide; microscopic transport mechanism; apparent permeability

1. Introduction

The shale reservoir matrix predominantly comprises micro-nano scale pores and fractures, wherein gas manifests unique transport mechanisms relative to those observed in larger-scale pores and fractures. Consequently, the advancement of micro-transport models for gases, particularly methane and carbon dioxide, assumes paramount importance in characterizing the macroscopic influences of multiple micro-transport mechanisms. Furthermore, this endeavor can lay a solid theoretical groundwork for the successful integration of carbon capture, utilization, and storage techniques within the realm of shale gas production.

At present, predecessors have considered multi-microscopic transport mechanisms, including bulk phase gas transport (slippage effect, Knudsen diffusion, and viscous flow), surface diffusion, real gas, adsorption layer, rarefied gas, and stress sensitivity [1–3], to establish the microscopic transport model of shale gas. Moreover, from a macroscopic application perspective, the apparent permeability is used to characterize the microscopic transport mechanism of gas. Thereafter, a multi-scale characterization model is established [4,5], which realizes the unity of the micro-mechanism and macro model.

Previous literature has laid the foundation for the establishment of multi-scale characterization models by clarifying the multiple micro-transport mechanisms of gas in micro-nano pores and fractures of shale matrix. For the micro-transport mechanism of bulk gas, Wu et al. proposed that the flow patterns in the shale matrix include continuous flow, slippage flow, and transitional flow [6]. Zhang pointed out that throughout the entire shale gas production cycle, all flow patterns, such as continuous flow, slippage flow, transition flow, and Knudsen diffusion, can occur in the nano pores of shale, and the transport mechanisms within different pore sizes are different [2]. Among them, continuous flow can be characterized by Darcy's law, which is widely used in conventional oil and gas reservoirs. The influence of the slippage effect can be characterized by modified permeability for slippage flow, and the modified coefficient of permeability can be obtained by fitting experimental data [7]. Knudsen diffusion is driven by pressure difference [8], which can be characterized by modifying the bulk gas permeability. Moreover, the microscopic transport mechanism of the adsorbed phase is characterized by surface diffusion [1]. The study of Sheng et al. showed that the concentration gradient was the driving force of surface diffusion, which can use the Langmuir equation and Fick's law to establish a surface diffusion model [9]. Zhang et al. pointed out that adsorbed gas molecules migrated under the action of the adsorption potential field, which can be characterized by combining the Maxwell–Stefan method with chemical potential [2]. In addition, Yin et al. proposed that the adsorption layer reduces the space for bulk gas transport; therefore, the influence of adsorption on gas transport in micro-nano pores is significant [10]. Previous studies considered the influence of adsorption layers by modifying permeability [1]. The research of Li et al. showed that the stress sensitivity effect is the main influencing factor of matrix apparent permeability, which controls the apparent permeability of shale matrix decrease with the increase in the permeability stress sensitivity coefficient [11]. Wu et al. proposed that the real gas effect has a significant impact on gas transport in shale reservoirs with high temperatures and pressures [12]. At present, the equation of state is employed to characterize the real gas effect in micro-transport models. The study of Wu and Chen illustrated that the real gas effect improves the gas transport ability in shale, which impact increases with increasing pressure, decreasing temperature, and decreasing pore size [1].

Based on the multiple microscopic transport mechanisms of gas, previous literature has considered different effects to establish the multi-scale mathematical representation model, which was employed to study the macro effect of multiple microscopic transport mechanisms. Ertekin et al. proposed a method to describe dynamic slip during gas transport by providing a slip factor related to pressure and composition, in which the viscous flow and Fick diffusion are considered [13]. Based on this, Clarkson et al. further employed gas desorption and established a flow model [14]. Besides, Javadpour proposed an apparent permeability formula, which characterizes Knudsen diffusion based on the interaction between gas molecules and their pore walls. This method improved the Darcy formula and described gas flow in mudstone micropores, which also illustrated that the smaller the pore size and the lower the pressure and temperature conditions, the more significant the impact of Knudsen diffusion on gas transport [15]. Considering slippage flow and Knudsen diffusion, Swami and Settari further established a numerical model for adsorbed gas and dissolved gas in micro-nano pores. The results indicated that changes in the matrix's apparent permeability should be considered in reservoir simulation [16]. Moreover, Sheng et al. established a transient capillary flow model by considering the mechanisms of intrapore diffusion, surface diffusion, viscous slippage flow, and gas desorption. Its result shows that the influence of surface diffusion increases with the decrease in pore size [9]. Li et al. established the apparent permeability model of gas by considering viscous flow, Knudsen diffusion, and surface diffusion. Their study pointed out that under the conditions of low pressure and high pressure, the influence of various factors on the apparent permeability showed different change laws [17]. Tian et al. considered slippage flow, bulk diffusion, and Knudsen diffusion by a weighted method, which is employed to study the contribution rate of each micro-transport mechanism to the total flow rate [18]. Huang et al. studied the contribution of surface diffusion, slippage flow, Knudsen diffusion, and viscous flow to the total gas flow in nano pores of shale by comprehensively considering the pore size and pressure, pore wall roughness, pore mechanics reaction, adsorption-induced expansion reaction, and influence of the adsorption layer [19]. Considering surface diffusion, effective stress, and gas adsorption, Song et al. established a gas transport model by weighted continuous flow and Knudsen diffusion to characterize the evolution mechanism of dynamic apparent permeability of the shale reservoir [20].

In summary, the previously proposed multi-scale characterization model for gas micro-transport takes into account the microscale effects and reservoir characteristics in a more comprehensive manner. However, given the diverse array of micro-transport mechanisms for gas, there is a need to enhance the gas transport model within the micro-nano pores of the shale matrix by considering factors such as temperature, pore structure, and material composition. In light of this, our study presents a novel approach that incorporates the temperature effect and the heterogeneity of micro-nano pores in the shale matrix. By integrating various phenomena including bulk phase gas transport (continuous flow, slippage flow, transition flow, and Knudsen diffusion), adsorption layer effect, surface diffusion effect, real gas effect, and stress sensitivity effect, we establish a multi-scale characterization model for gas micro-transport mechanisms in shale matrix micro-nano pores. This model not only serves as a theoretical foundation but also paves the way for the practical application of gas micro-transport mechanisms on a macroscopic scale.

The main work of this study is as follows: First, clarify the multiple microscopic transport mechanism of gas in micro-nano pores and fractures. The apparent permeability is used to realize the multi-scale characterization of the microscopic transport mechanism. The microscopic transport model of gas in micro-nano pores and fractures of the shale matrix is established. Secondly, the accuracy of the model is verified by experimental data. Finally, based on the microscopic transport model of gas proposed in this paper, the influencing factors of the macro effect of multiple microscopic transport mechanisms of gas are quantitatively analyzed.

2. Mechanism and Model

2.1. Transport Mechanism of Gas

2.1.1. Bulk Gas

Due to the size of micro-nano pores and fractures being small, the flow pattern of bulk gas in it is different from the Darcy flow. In this paper, the Knudsen number is used as the criterion for judging the flow pattern of bulk gas in the micro-nano pores and fractures. Thereafter, the correction function corrected slippage flow is weighted with Knudsen diffusion to establish a double slippage model, which can be used to characterize the continuous flow, slippage flow, transition flow, and Knudsen diffusion of methane and carbon dioxide.

Flow Pattern

The Knudsen number (Kn) is the basis for determining the flow pattern of the gas. The Kn can be calculated by Equation (1). When $Kn \leq 0.001$, the flow of gas in porous media is continuous, which can be characterized by Darcy's law. When $0.001 < Kn \leq 0.1$, the slippage flow appears. Although viscous flow still dominates, the slippage effect should be characterized by the correction of Darcy's formula by slippage theory. When $0.1 < Kn \leq 10$, it is the transition flow, which is when the probability of the collision between fluid–fluid molecules is equivalent to the collision between fluid–wall molecules. At this stage, the continuous flow assumption is not valid. When $Kn > 10$, since the average free path of molecules is larger than the pore size, the collision between fluid molecules and the wall is dominant. Therefore, this flow stage is Knudsen diffusion.

$$Kn = \frac{\lambda}{L} \tag{1}$$

where Kn is the Knudsen number, dimensionless; λ is the average free path of the molecule, m; L is the characteristic length, which is equal to the pore diameter, m.

Considering the real gas effect, the average free path of molecules in Formula (1) can be calculated by Formula (2) [21].

$$\lambda = \frac{\mu}{P}\sqrt{\frac{\pi ZRT}{2M}} \tag{2}$$

where μ is the gas viscosity, Pa·s; P is the pressure, Pa; M is the relative molecular mass, kg/mol; Z is the gas compressibility factor, dimensionless; R is the gas constant, m^3·Pa/(K·mol); and T is the temperature, K.

Continuous Flow

Continuous flow can be described by Darcy's formula, which can be characterized as the following:

$$J_{continuous} = -\rho\frac{K}{\mu}\nabla P \tag{3}$$

where $J_{continuous}$ is the mass flux, kg/(m^2·s); ρ is the gas density, kg/m^3; and K is the permeability, 10^{15} mD.

Slippage Flow

For the slippage flow, the assumption of the continuous medium fails and the rarefied gas effect becomes apparent. Therefore, the fluid velocity at the boundary is no longer zero, which results in the slippage effect, also known as the Klinkenberg effect. In this paper, based on the Beskok–Karniadakis model, the correction function related to the Knudsen number is employed to correct the continuous flow formula. By correcting Darcy's formula, the slippage flow in micropores and fractures is described. The correction equation can be expressed by Equation (4) as the following:

$$K_{slip} = K_o f(Kn) \tag{4}$$

For the circular flow section of the micropore and the rectangular flow section of the microfracture, the $f(Kn)$ correction functions are shown in Formulas (5) and (6), respectively. The characteristic lengths are pore diameter and fracture width.

$$f(Kn) = (1 + 2\alpha Kn)\left(1 + \frac{8Kn}{1 - 2bKn}\right) \tag{5}$$

$$f(Kn) = (1 + \alpha Kn)\left(1 + \frac{6Kn}{1 - bKn}\right) \tag{6}$$

where α is the rarefaction coefficient, dimensionless; b is the empirical slippage coefficient; and $b = -1$ represents that the slippage flow is related to the flow channel and is independent of the gas type, dimensionless.

$$\alpha = \frac{128}{3\pi^2(1 - 4b)}\arctan\left(4Kn^{0.4}\right) \tag{7}$$

Darcy's formula can be written as the following:

$$v = -\frac{K}{\mu}\frac{dP}{dx} \tag{8}$$

Substituting Equations (4) and (5) into Equation (8), the flow rate of slippage flow in the micropore can be expressed as Equation (9) as the following:

$$v_{slip} = -\frac{K_o}{\mu}(1+2\alpha Kn)\left(1+\frac{8Kn}{1-2bKn}\right)\frac{dP}{dx} \tag{9}$$

For the micropore,

$$v = -\frac{1}{8}\frac{r^2}{\mu}\frac{dP}{dx} \tag{10}$$

Therefore, $K_o = \frac{r^2}{8}$, and the Equation (9) can be written as the following:

$$v_{slip} = -\frac{r^2}{8\mu}(1+2\alpha Kn)\left(1+\frac{8Kn}{1-2bKn}\right)\frac{dP}{dx} \tag{11}$$

Similarly, substituting Equations (4) and (6) into Equation (8), the flow rate of slippage flow in the microfracture can be expressed as Equation (12), as the following:

$$v_{slip} = -\frac{K_o}{\mu}(1+\alpha Kn)\left(1+\frac{6Kn}{1-bKn}\right)\frac{dP}{dx} \tag{12}$$

For the microfracture,

$$v = -\frac{1}{12}\frac{h^2}{\mu}\frac{dP}{dx} \tag{13}$$

Therefore, $K_o = \frac{h^2}{12}$, and the Equation (12) can be written as the following:

$$v_{slip} = -\frac{h^2}{12\mu}(1+\alpha Kn)\left(1+\frac{6Kn}{1-bKn}\right)\frac{dP}{dx} \tag{14}$$

In summary, the mass flux of micropores can be expressed by Equation (15). The mass flux of microfractures can be expressed by Equation (16).

$$\begin{aligned} J_{slip} &= -\rho\frac{K_o}{\mu}\frac{\phi}{\tau}(1+2\alpha Kn)\left(1+\frac{8Kn}{1-2bKn}\right)\frac{dP}{dx} \\ &= -\frac{r^2}{8\mu}\frac{\phi}{\tau}\frac{PM}{ZRT}(1+2\alpha Kn)\left(1+\frac{8Kn}{1-2bKn}\right)\frac{dP}{dx} \end{aligned} \tag{15}$$

$$\begin{aligned} J_{slip} &= -\rho\frac{K_o}{\mu}\frac{\phi}{\tau}(1+\alpha Kn)\left(1+\frac{6Kn}{1-bKn}\right)\frac{dP}{dx} \\ &= -\frac{h^2}{12\mu}\frac{\phi}{\tau}\frac{PM}{ZRT}(1+\alpha Kn)\left(1+\frac{6Kn}{1-bKn}\right)\frac{dP}{dx} \end{aligned} \tag{16}$$

where J_{slip} is the slippage flux, kg/(m^2·s); ϕ is porosity, dimensionless; and τ is tortuosity, dimensionless.

Knudsen Diffusion

For Knudsen diffusion, the average free path of gas molecules is much larger than the pore size. The collision of gas–wall molecules is dominant and the collision of gas–gas molecules can be ignored. Therefore, Knudsen diffusion has a constant diffusion coefficient, which can neglect viscosity effects [22]. Knudsen diffusion can be characterized by Equation (17).

$$\frac{q\rho}{\phi AM} = D_k\frac{\Delta P}{RTL} \tag{17}$$

where q is the flow rate, m^3/s; A is the overcurrent area, m^2; and D_k is the Knudsen diffusion coefficient, m^2/s.

The Knudsen diffusion coefficient can be expressed by Equation (18).

$$D_k = \frac{d}{3}\sqrt{\frac{8RT}{\pi M}} \tag{18}$$

where d is the pore diameter, m.

Considering the average velocity of real gas molecules, $\overline{u} = \sqrt{\frac{8ZRT}{\pi M}}$, thus, the Knudsen diffusion coefficient can be expressed by Equation (19).

$$D_k = \frac{d}{3}\sqrt{\frac{8ZRT}{\pi M}} \tag{19}$$

Substitute Equation (19) into Equation (17), and the mass flux of the Knudsen diffusion can be obtained.

$$J_{Knudsen} = \frac{q}{A}\rho = -\frac{MD_k}{RT}\frac{dP}{dx} = -\frac{d}{3}\sqrt{\frac{8ZM}{\pi RT}}\frac{dP}{dx} = -\frac{2}{3}r\sqrt{\frac{8ZM}{\pi RT}}\frac{dP}{dx} \tag{20}$$

where $J_{Knudsen}$ is the Knudsen diffusion mass flux of the real gas, $kg/(m^2{\cdot}s)$.

In the condition of the relatively rough pore wall, the Knudsen diffusion is slowed down due to the longer time that the gas stays near the pore wall. Therefore, the influence of pore wall roughness on Knudsen diffusion needs to be considered [19]. The geometric shape of pores in porous media is complex, and pores are connected to each other [23–25]. Moreover, the actual path of gas transmission is longer than the direct path (tortuosity), and the ratio of porosity and tortuosity should be used to correct the Knudsen diffusion coefficient [26]. Therefore, the mass flux of Knudsen diffusion can be written as:

$$J_{Knudsen} = -\frac{d}{3}\frac{\phi}{\tau}\delta^{D_f-2}\sqrt{\frac{8ZM}{\pi RT}}\frac{dP}{dx} = -\frac{2}{3}r\frac{\phi}{\tau}\delta^{D_f-2}\sqrt{\frac{8ZM}{\pi RT}}\frac{dP}{dx} \tag{21}$$

where δ is the ratio of molecular diameter to local pore diameter, dimensionless; and D_f is the fractal dimension of pore wall, dimensionless.

Transition Flow

Transition flow is the transition flow state between slippage flow and Knudsen diffusion. This paper combined slippage flow with Knudsen diffusion by weight coefficients to characterize the transition flow, as shown in Equation (22).

$$J_{Trans} = \omega_{Knudsen}J_{Knudsen} + \omega_{Slip}J_{Slip} \tag{22}$$

where $\omega_{Knudsen}$ is the Knudsen diffusion weight coefficient, dimensionless; ω_{Slip} is the weight coefficient of slippage flow, dimensionless; and J_{Trans} is the mass flux of the real gas transition flow, $kg/(m^2{\cdot}s)$.

Previous scholars have proposed multiple weight coefficient formulas through derivation or fitting [8,27,28]. This study uses the corrected model from Shi et al. [28] to calculate the weight coefficients.

$$\omega_{slip} = \frac{1}{1 + (Kn/Kn_{0.5})^{n-1}} \tag{23}$$

$$\omega_{Knudsen} = 1 - \omega_{slip} \tag{24}$$

where $Kn_{0.5}$ is a constant, taken as 1.03; and n is a constant, taken as 3.689.

2.1.2. Adsorption Gas

The adsorbed state is one of the important occurrence states of gas in micro-nano pores of the shale matrix. When gas is adsorbed on the pore wall, the gas molecules in the adsorption layer migrate under the action of the adsorption potential field, which is surface diffusion [19]. In this paper, the Maxwell–Stefan method is used to characterize surface diffusion. This method assumes that the driving force of surface diffusion is the chemical potential gradient, and its surface diffusion flux can be written as the following:

$$J_{surface} = -L_m C_s \nabla \psi \tag{25}$$

The parameters in Equation (25) can be calculated by the following equations:

$$L_m = \frac{D_s}{RT} \tag{26}$$

$$\psi = \psi_o + RT\ln(P) \tag{27}$$

$$C_s = \frac{4M\theta}{\pi d_m^3 N_A} \tag{28}$$

where $J_{surface}$ is the surface diffusion mass flux, kg/(m^2·s); L_m is the gas flow fluidity, mol·s/kg; C_s is the concentration of adsorbed gas, kg/m^3; ψ is chemical potential, J/mol; D_s is the surface diffusion coefficient, m^2/s; d_m is the diameter of the gas molecule, m; and N_A is the Avogadro constant, mol^{-1}.

Substitute Equations (26)–(28) into Equation (25), and the surface diffusion flux can be obtained and expressed as Equation (29).

$$J_{surface} = -\frac{D_s C_{sc}}{P}\frac{dP}{dx} \tag{29}$$

The surface diffusion coefficient in Equation (29) can be calculated by Equation (30). The parameters can be calculated by Equations (31)–(33), respectively.

$$D_s = D_{so}\frac{(1-\theta) + \frac{\kappa}{2}\theta(2-\theta) + [H(1-\kappa)](1-\kappa)\frac{\kappa}{2}\theta^2}{\left(1-\theta+\frac{\kappa}{2}\theta\right)^2} \tag{30}$$

$$D_{so} = 8.29 \times 10^{-7} T^{0.5} \exp\left(-\frac{\Delta H^{0.8}}{RT}\right) \tag{31}$$

$$H(1-\kappa) = \begin{cases} 0, & \kappa \geq 1 \\ 1, & 0 \leq \kappa < 1 \end{cases} \tag{32}$$

$$\kappa = \frac{\kappa_b}{\kappa_m} \tag{33}$$

where $H(1-\kappa)$ is the Heaviside function, dimensionless; κ is the blocking coefficient of gas molecules on the pore wall, dimensionless; κ_b is the blocking velocity coefficient of gas molecules on the pore wall, m/s; κ_m is the forward velocity coefficient of gas molecules on the pore wall, m/s; D_{so} is the surface diffusion coefficient with gas coverage of "0", m^2/s; and ΔH is the isothermal adsorption heat of surface coverage rate θ = 0, J/mol.

2.2. Changing Mechanism of Reservoir

During the development of shale reservoirs, there is stress sensitivity, matrix shrinkage effects, and adsorption layer, which make the size of matrix pores change with pressure [19].

2.2.1. Stress Sensitive Effect

Shale reservoir has strong stress sensitivity. With the decrease in reservoir pressure, the effective stress of the formation gradually increases. Its macro performance is the reduction of matrix permeability and porosity. This research employed exponential formulas to describe the stress sensitivity of permeability and porosity [29].

$$K = K_o \exp\left[3c_p(P - P_o)\right] \tag{34}$$

$$\phi = \phi_o \exp\left[c_p(P - P_o)\right] \tag{35}$$

where K_o is the initial absolute permeability, mD; P_o is the initial pressure, Pa; c_p is the pore compression coefficient, Pa^{-1}; and ϕ_o is the initial porosity, dimensionless.

The relationship between porosity, permeability, tortuosity, and pore radius is obtained from Carman–Kozeny formula, which can be expressed as Equation (36).

$$r = 2\sqrt{2\tau}\sqrt{\frac{K}{\phi}} \tag{36}$$

Substitute Equations (34) and (35) into Equation (36), and Equation (37) can be calculated, which is used to characterize the relationship between reservoir pressure and pore radius.

$$r = r_o\sqrt{\exp\left[2c_p(P - P_o)\right]} \tag{37}$$

where r_o is the initial pore throat radius, m.

2.2.2. Matrix Shrinkage Effect

As the pressure of the shale reservoir decreases, the gas in the matrix pores will undergo desorption and be released. The matrix contracts due to the pressure decrease in the pore. Therefore, the volume of microfractures is increasing, which results in the apparent permeability of the matrix increasing. Moreover, previous studies show that matrix shrinkage has a significant impact on reservoir permeability [30,31]. This paper uses the formula proposed by Wang et al. to characterize the matrix shrinkage effect [32].

$$\omega_{swelling} = \frac{K}{K_o} = \left(1 - \frac{3}{\phi_o}\frac{\varepsilon_L P_L (P - P_o)}{(P + P_L)(P_o + P_L)}\right)^3 \tag{38}$$

where $\omega_{swelling}$ is the matrix shrinkage coefficient, dimensionless; ε_L is Langmuir strain, dimensionless; and P_L is the Langmuir pressure, Pa.

2.2.3. Effect of Adsorption Layer

Due to the adsorption gas in the shale matrix occupying the pore space, the effective pore size for bulk gas transport is reduced. As the pressure of shale reservoirs decreases, gas gradually desorbs, and the effective pore size for gas transmission increases [33]. Therefore, the transport of gas in matrix pores is affected by gas adsorption and desorption. This paper considers the thickness of one methane molecule as the maximum adsorption layer thickness. The ratio of adsorption to maximum adsorption is set as the coverage, to approximately calculate the pore size of shale matrix under the influence of the adsorption layer. For micropores and microfractures, the effective pore size and fracture width are expressed as Equation (41) and Equation (42), respectively.

$$\theta = \frac{N}{N_L} \tag{39}$$

$$r_{ad} = d_m\theta \tag{40}$$

$$r_e = r - d_m\theta \tag{41}$$

$$h_e = h - 2d_m\theta \tag{42}$$

where θ is the coverage of adsorbed gas, dimensionless; N is the adsorption amount, mmol/g; N_L is the maximum adsorption amount, mmol/g; r_{ad} is the thickness of the adsorption layer, m; d_m is the diameter of methane molecules, m; r_e is the effective radius of the pore, m; and h_e is the effective width of the fracture, m.

2.3. Transport Model of Gas

This article employed the apparent permeability model to characterize the mass flux model with multiple transmission mechanisms. The relationship between permeability and mass flux can be expressed in Equation (43).

$$K = \frac{J\mu ZRT}{MP\frac{dP}{dx}} \tag{43}$$

For slippage flow, the apparent permeability model in the micropore can be obtained from Equations (15) and (43). The apparent permeability model in the microfracture can be obtained from Equations (16) and (43). Therefore, the apparent permeability models of micropore and microfracture are expressed as Equation (44) and Equation (45), respectively.

$$K_{slip} = \frac{r^2}{8}\frac{\phi}{\tau}(1 + 2\alpha Kn)\left(1 + \frac{8Kn}{1 - 2bKn}\right) \tag{44}$$

$$K_{slip} = \frac{h^2}{12}\frac{\phi}{\tau}(1 + \alpha Kn)\left(1 + \frac{6Kn}{1 - bKn}\right) \tag{45}$$

For Knudsen diffusion, the apparent permeability model obtained from Equations (20) and (43) is expressed as follows:

$$K_{Knudsen} = \frac{d}{3}\frac{\phi}{\tau}\delta^{D_f-2}\frac{\mu}{P}\sqrt{\frac{8ZRT}{\pi M}} \tag{46}$$

For surface diffusion, according to Equations (29) and (43), the apparent permeability model of surface diffusion is written in Equation (47) as the following:

$$K_{surface} = \frac{D_sC_{sc}}{P}\frac{\phi}{\tau}\frac{\mu ZRT}{PM} \tag{47}$$

Based on the above models, apparent permeability models for micropores and microfractures in shale reservoirs are established, respectively. For organic micropores, the transport of bulk gas and adsorption gas is considered, and apparent permeability is expressed by Equation (48). In inorganic micropores, bulk gas transport is considered only. Thus, its apparent permeability is expressed as Equation (49).

$$\begin{aligned} K_{pore}^{organic} &= \omega_{slip}K_{slip} + \omega_{Knudsen}K_{Knudsen} + K_{surface} \\ &= \omega_{slip}\tfrac{r^2}{8}\tfrac{\phi}{\tau}(1 + 2\alpha Kn)\left(1 + \tfrac{8Kn}{1-2bKn}\right) + \\ &\omega_{Knudsen}\tfrac{2}{3}r\tfrac{\phi}{\tau}\delta^{D_f-2}\tfrac{\mu}{P}\sqrt{\tfrac{8ZRT}{\pi M}} + \tfrac{D_sC_{sc}}{P}\tfrac{\phi}{\tau}\tfrac{\mu}{\rho_g} \end{aligned} \tag{48}$$

$$\begin{aligned} K_{pore}^{inorganic} &= \omega_{slip}K_{slip} + \omega_{Knudsen}K_{Knudsen} \\ &= \omega_{slip}\tfrac{r^2}{8}\tfrac{\phi}{\tau}(1 + 2\alpha Kn)\left(1 + \tfrac{8Kn}{1-2bKn}\right) + \\ &\omega_{Knudsen}\tfrac{2}{3}r\tfrac{\phi}{\tau}\delta^{D_f-2}\tfrac{\mu}{P}\sqrt{\tfrac{8ZRT}{\pi M}} \end{aligned} \tag{49}$$

For microfractures, the transport of bulk gas and adsorption gas is considered to set up the apparent permeability model in organic microfractures. This can be taken to be represented by Equation (50). However, in inorganic microfractures, bulk gas transport is considered, and the apparent permeability model is expressed in Equation (51).

$$\begin{aligned} K_{fracture}^{organic} &= \omega_{slip}K_{slip} + \omega_{Knudsen}K_{Knudsen} \\ &= \omega_{slip}\frac{h^2}{12}\frac{\phi}{\tau}(1+\alpha Kn)\left(1+\frac{6Kn}{1-bKn}\right)+ \\ & \omega_{Knudsen}\frac{2}{3}h\frac{\phi}{\tau}\delta^{D_f-2}\frac{\mu}{P}\sqrt{\frac{8ZRT}{\pi M}} + \frac{D_sC_{sc}}{P}\frac{\phi}{\tau}\frac{\mu}{\rho_g} \end{aligned} \tag{50}$$

$$\begin{aligned} K_{fracture}^{inorganic} &= \omega_{slip}K_{slip} + \omega_{Knudsen}K_{Knudsen} \\ &= \omega_{slip}\frac{h^2}{12}\frac{\phi}{\tau}(1+\alpha Kn)\left(1+\frac{6Kn}{1-bKn}\right)+ \\ & \omega_{Knudsen}\frac{2}{3}h\frac{\phi}{\tau}\delta^{D_f-2}\frac{\mu}{P}\sqrt{\frac{8ZRT}{\pi M}} \end{aligned} \tag{51}$$

This paper takes the proportion of micropores and microfractures in the organic and inorganic matter to the total volume of pores and fractures as the weight coefficient. Meanwhile, considering the matrix shrinkage effect, the apparent permeability model of the shale reservoir matrix is established.

$$\begin{aligned} K_{app} &= \left[\gamma\left(\alpha K_{pore}^{organic} + (1-\alpha)K_{pore}^{inorganic}\right)+ \right. \\ & \left. (1-\gamma)\left(\beta K_{fracture}^{organic} + (1-\beta)K_{fracture}^{inorganic}\right)\right]\omega_{swelling} \end{aligned} \tag{52}$$

where α is the proportion of organic matter pore volume to total pore volume, dimensionless; β is the proportion of organic matter fracture volume to total fracture volume, dimensionless; and γ is the proportion of pore volume to total pore volume, dimensionless.

3. Model Validation

To verify the accuracy of the apparent permeability model of the shale matrix established in this paper, experimental data from Shi et al. [34] were used to validate the model. The fitting parameters used for validation are shown in Table 1, and the other basic parameters are shown in Table 2.

Table 1. The fitting parameters of the apparent permeability model.

Fitting Parameters (Experiment 1)	Values	Fitting Parameters (Experiment 2)	Values
Initial pressure P_o (MPa)	40	Initial pressure P_o (MPa)	25
Porosity ϕ (f)	0.015	Porosity ϕ (f)	0.05
Rock compressibility factor C_p (MPa^{-1})	0.013	Rock compressibility factor C_p (MPa^{-1})	0.012

Table 2. Calculation parameters for the bulk gas transport model.

Parameters	Values	Parameters	Values
Temperature T (K)	300	Fluid viscosity μ (mPa·s)	0.0175
Porosity ϕ (f)	0.05	Molecular molar mass M (kg/mol)	0.028
Tortuosity τ (f)	4.3	The ratio of molecular diameter to local pore diameter σ (f)	0.5
Fracture width h (nm)	50	Fractal dimension of pore wall D_f (f)	2.5

As the results show in Figure 1, the variation pattern of matrix apparent permeability with pressure described by this model is consistent with the experimental results. Therefore, the shale matrix apparent permeability model established in this paper based on the multiple microscopic transmission mechanisms can be used to characterize the multiple transmission mechanisms of shale gas. Moreover, the change rule of reservoir matrix permeability can be studied by this model, which has practicability.

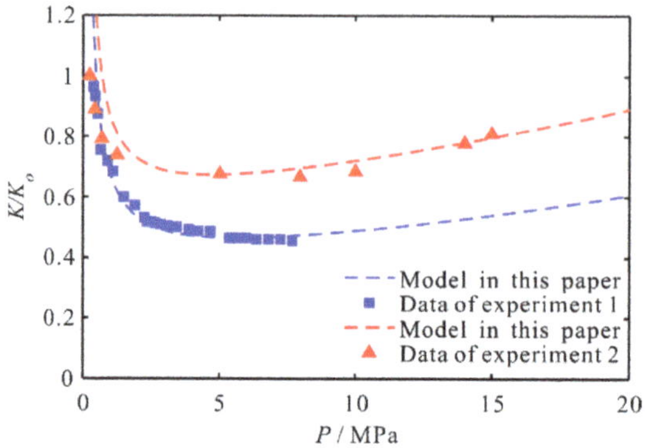

Figure 1. Comparison of permeability between experimental data and the model presented in this paper.

4. Analysis and Discussion

4.1. *Influence for Knudsen Number*

According to the general conditions of shale gas reservoirs, the pressure and pore size distribution ranges are 1~50 MPa and 2~1000 nm, respectively [1]. This paper calculates the Knudsen numbers with pore sizes of 1 nm, 10 nm, 100 nm, and 1000 nm at temperatures of 333 K and 363 K, with pressures of 0.1 MPa to 80 MPa, respectively, as shown in Figure 2. It can be seen that during the development of the shale gas reservoir, the flow pattern of shale gas can include all the flow patterns from Knudsen diffusion to continuous flow. Moreover, as the aperture increases, the Knudsen number decreases. As the pressure increases, the Knudsen number decreases. As the temperature increases, the Knudsen number increases (see Figure 2).

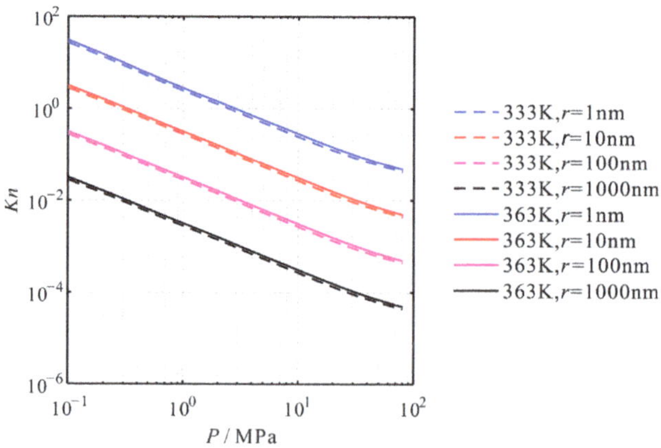

Figure 2. Relationship between Knudsen number and pressure, temperature, and pore size.

Considering the real gas effect in the micro-nano pores of the shale, the influence of micro-nano pores on the physical properties of methane is characterized by the compressibility factor in Equation (2). As shown in Figure 3, considering the compressibility factor, the Knudsen number decreases. In small apertures, the decrease in the Knudsen number is more pronounced.

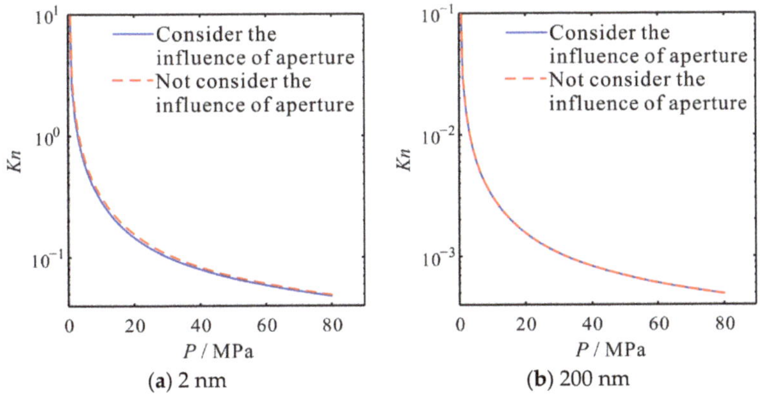

Figure 3. Distribution range of Knudsen number for different pore sizes and pressures.

4.2. Influence of Pressure

Under the same conditions, using pressure as a variable, the micro-transport mechanisms and the changes in matrix apparent permeability are studied. As shown in Figure 4a, with pressure decreasing from 50 MPa to 0 MPa, the apparent permeability of the surface diffusion and the Knudsen diffusion increases by two and eight orders of magnitude, respectively. However, the apparent permeability of slippage flow decreases by two orders of magnitude. Thereafter, the contributions of the three transport mechanisms to the apparent permeability of the shale matrix are analyzed (see Figure 4b). As the pressure decreases from 50 MPa to 0 MPa, the contribution proportion of slippage flow decreases from 1 to 0. The contribution proportion of Knudsen diffusion increases from 0 to 0.72. The contribution proportion of surface diffusion increases first and then decreases, with the largest contribution proportion at 3.6 MPa, 0.49. Moreover, when the pressure is low, the changing of the contribution proportion for the three transport mechanisms to the apparent permeability can be seen more significantly. It can be seen that pressure changes have a significant impact on the values and contribution proportions of the apparent permeability of the three transmission mechanisms.

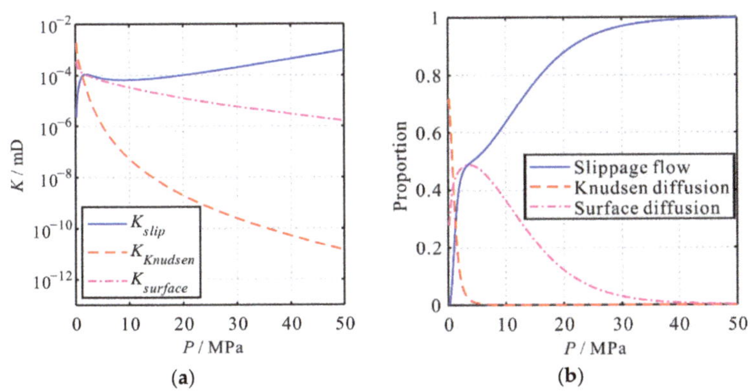

Figure 4. The influence of various transport mechanisms on the apparent permeability of shale matrix with changes in pressure. (**a**) The apparent permeability. (**b**) The contributions for the apparent permeability.

As shown in Figure 5, as pressure decreases, the apparent permeability of the matrix shows a trend of first decreasing and then increasing. Its value varies from 9.2×10^{-4} mD at the initial 50 MPa, reduced to 1.1×10^{-4} mD at 14.5 MPa. Then, values rise to 4.5×10^{-3} mD at 0 MPa. Based on the relationship between the transport mechanism, the apparent permeability, and pressure, it can be concluded that as pressure decreases, the apparent permeability of the matrix decreases, since the apparent permeability of the slippage flow decreases. With the further decrease in pressure, the apparent permeability of the shale matrix increases due to the increase in apparent permeability contributed by Knudsen diffusion and surface diffusion.

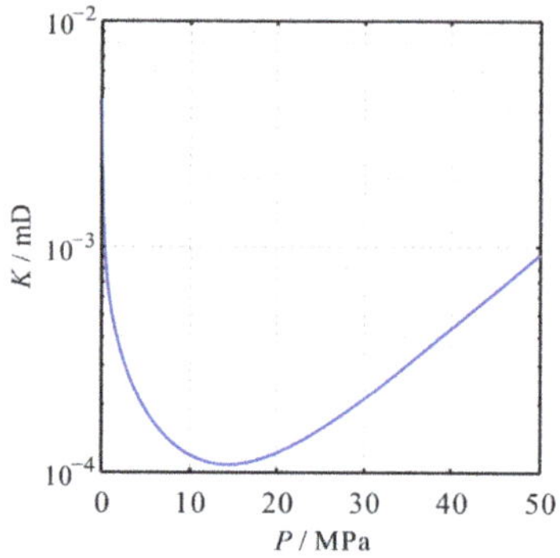

Figure 5. The variation trend of apparent permeability of shale matrix with pressure.

4.3. *Influence of Temperature*

Under the same conditions, using temperature as a variable, the micro-transport mechanisms and the changes in matrix apparent permeability are studied. As shown in Figure 6a, with the temperature increase from 300 K to 400 K, the apparent permeability of the slippage flow, surface diffusion, and the Knudsen diffusion increase by 2.49%, 106.92%, and 89.78%, respectively. Thereafter, the contributions of the three transport mechanisms to the apparent permeability of the shale matrix are analyzed (see Figure 6b). As the temperature increases from 300 K to 400 K, the contribution proportion of slippage flow decreases from 0.98 to 0.96. The contribution proportion of surface diffusion increases from 0.02 to 0.04. The contribution proportion of Knudsen diffusion is small, although it increases. It can be seen that temperature changes have a significant impact on the values of the apparent permeability of the three transmission mechanisms. The impact on the contribution proportions of the apparent permeability is small.

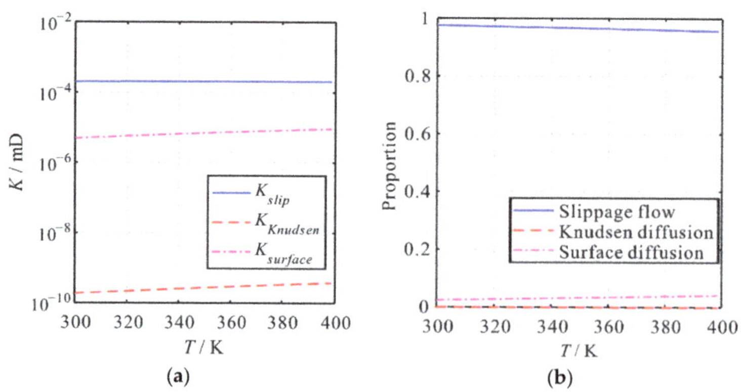

Figure 6. The influence of various transport mechanisms on the apparent permeability of shale matrix with changes in temperature. (**a**) The apparent permeability. (**b**) The contributions for the apparent permeability.

As shown in Figure 7, as temperature increases, the apparent permeability of the matrix increases. Its value varies from 2.075×10^{-4} mD at the initial 300 K, raised to 2.170×10^{-4} mD at 400 K. Based on the relationship between the transport mechanism, the apparent permeability, and temperature, it can be concluded that as temperature increases, the apparent permeability of the matrix increases. It is because the apparent permeability of slippage flow, Knudsen diffusion, and surface diffusion all increase.

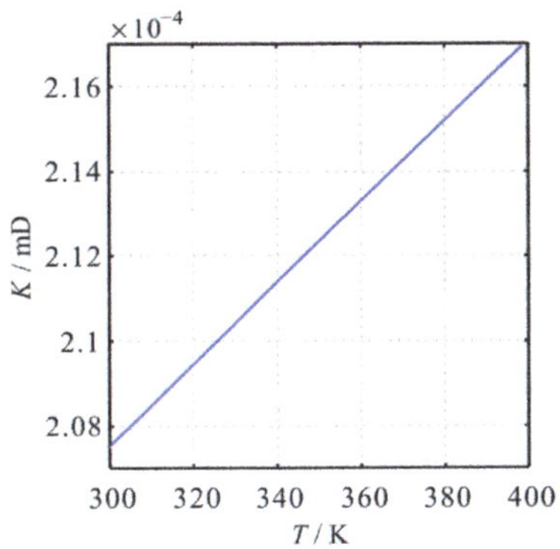

Figure 7. The variation trend of apparent permeability of shale matrix with temperature.

4.4. Influence of Pore Size

Under the same conditions, using pore size as a variable, the micro-transport mechanisms and the changes in matrix apparent permeability are studied. As shown in Figure 8a, with pore size increasing from 1×10^{-9} m to 1×10^{-6} m, the apparent permeability of the surface diffusion and the Knudsen diffusion decreases by three and six orders of magnitude, respectively. However, the apparent permeability of slippage flow increases by six orders of magnitude. Thereafter, the contributions of the three transport mechanisms to the apparent permeability of the shale matrix are analyzed (see Figure 8b). As the pore size increases from 1×10^{-9} m to 1×10^{-6} m, the contribution proportion of slippage flow increases from 0.015 to 1. The contribution proportion of surface diffusion decreases from 0.985 to 0. The contribution proportion of Knudsen diffusion increases first and then decreases, although it is small. It can be seen that pore size changes have a significant impact on the values and contribution proportions of the apparent permeability of the three transmission mechanisms.

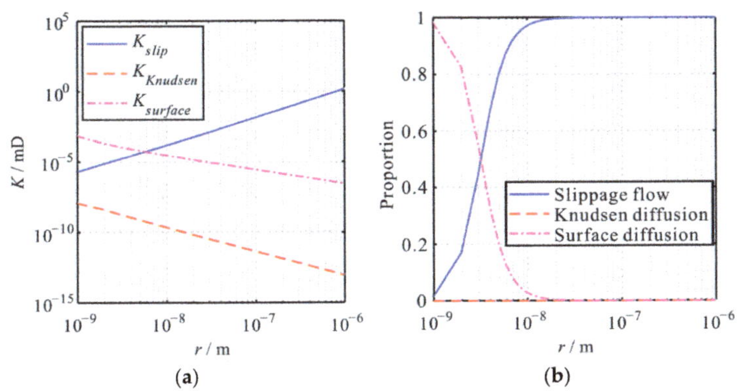

Figure 8. The influence of various transport mechanisms on the apparent permeability of shale matrix with changes in the pore size. (**a**) The apparent permeability. (**b**) The contributions for the apparent permeability.

As shown in Figure 9, as pore size increases, the apparent permeability of the matrix shows a trend of first decreasing and then increasing. Its value varies from 1.66×10^{-4} mD at the initial 1×10^{-9} m, reduced to 1.90×10^{-4} mD at 1×10^{-6} m. Based on the relationship between the transport mechanism, the apparent permeability, and pore size, it can be concluded that as pore size increases from 1 nm to 3 nm, the apparent permeability of the matrix decreases, since the apparent permeability of the surface diffusion decreases. With the further increase in pore size, the apparent permeability of the shale matrix increases due to the increase in apparent permeability contributed by slippage flow.

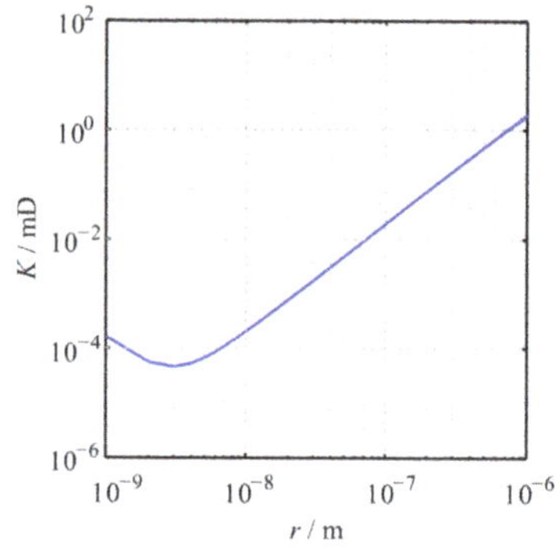

Figure 9. The variation trend of apparent permeability of shale matrix with pore size.

4.5. Influence of Tortuosity

Under the same conditions, using tortuosity as a variable, the micro-transport mechanisms and the changes in matrix apparent permeability are studied. As shown in Figure 10a, with tortuosity increasing from one to five, the apparent permeability of the slippage flow, surface diffusion, and Knudsen diffusion decrease by 80%, respectively. Thereafter, the contributions of the three transport mechanisms to the apparent permeability of the shale matrix are analyzed (see Figure 10b). As the tortuosity increases from one to five, the contribution proportions of slippage flow, surface diffusion, and Knudsen diffusion are maintained at 0.97, 0.03, and 0, respectively. It can be seen that tortuosity changes have a significant impact on the values of the apparent permeability of the three transmission mechanisms. However, the tortuosity change has no impact on the contribution proportions of the three transmission mechanisms.

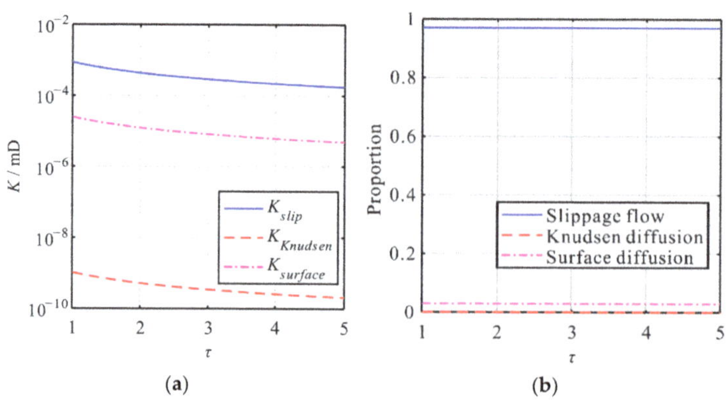

Figure 10. The influence of various transport mechanisms on the apparent permeability of shale matrix with changes in the tortuosity. (**a**) The apparent permeability. (**b**) The contributions for the apparent permeability.

As shown in Figure 11, as tortuosity increases, the apparent permeability of the matrix shows a trend of increasing. Its value varies from 9.06×10^{-4} mD at the initial $\tau = 1$, reduced to 1.81×10^{-4} mD at $\tau = 5$. Based on the relationship between the transport mechanism, the apparent permeability, and tortuosity, it can be concluded that as tortuosity increases, the apparent permeability of the matrix decreases, since the apparent permeability of the slippage flow, surface diffusion, and Knudsen diffusion all decrease. Moreover, the contribution proportion of three micro-transport mechanisms to the apparent permeability is invariant.

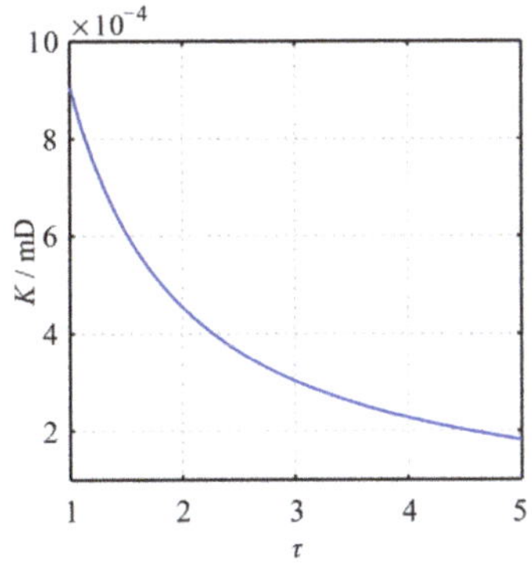

Figure 11. The variation trend of apparent permeability of shale matrix with the tortuosity.

4.6. Influence of Total Organic Carbon (TOC)

Under the same conditions, using TOC as a variable, the micro-transport mechanisms and the changes in matrix apparent permeability are studied. As shown in Figure 12a, with TOC increasing from zero to one, the apparent permeability of the slippage flow and the Knudsen diffusion is unchanged. The apparent permeability of the surface diffusion is increasing. Thereafter, the contributions of the three transport mechanisms to the apparent permeability of the shale matrix are analyzed (see Figure 12b). As the TOC increases from zero to one, the contribution proportions of slippage flow are reduced from 1 to 0.875. The contribution proportions of surface diffusion increased from 0 to 0.125. However, as the TOC increases, the contribution proportions of the Knudsen diffusion decrease gradually, which are always small.

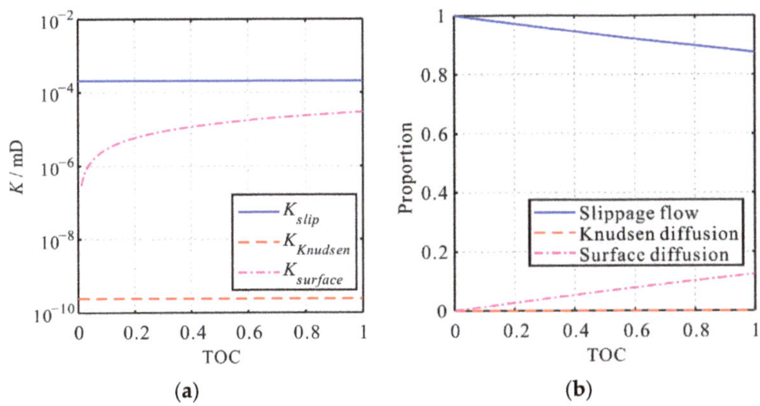

Figure 12. The influence of various transport mechanisms on the apparent permeability of shale matrix with changes in the TOC. (**a**) The apparent permeability. (**b**) The contributions for the apparent permeability.

As shown in Figure 13, as TOC increases, the apparent permeability of the matrix increases. Its value varies from 2.05×10^{-4} mD at the TOC = 0, raised to 2.34×10^{-4} mD at the TOC = 1. Based on the relationship between the transport mechanism, the apparent permeability, and TOC, it can be concluded that as TOC increases, the apparent permeability of the matrix increases. It is because the apparent permeability of surface diffusion increases.

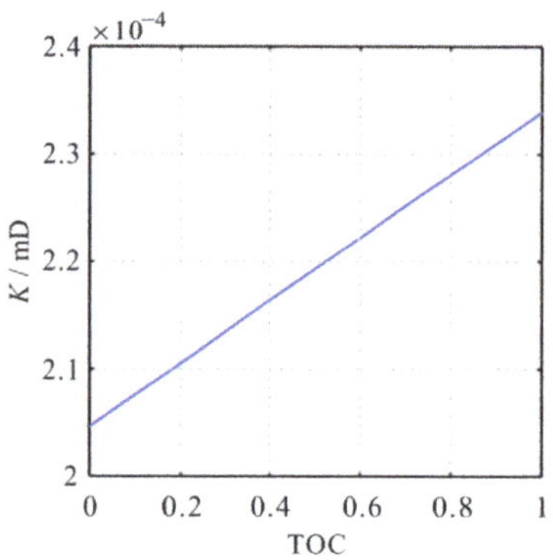

Figure 13. The variation trend of apparent permeability of shale matrix with the TOC.

4.7. Combined Influence

Considering the actual situation of the comprehensive action of multiple factors in the development of shale gas, the microscopic transport mechanism, the variation law of matrix apparent permeability in different pore sizes under different temperatures, and pressure conditions are analyzed.

As Figure 14 shows, the larger the pore size, the larger the apparent permeability of the shale matrix under the condition of constant temperature. With the decrease in pressure, the apparent permeability of the shale matrix with different pore sizes decreases first and then increases. With the increase in temperature, the apparent permeability of the matrix increases, and its variation amplitude increases with the decrease in pore size and pressure (see Figure 15).

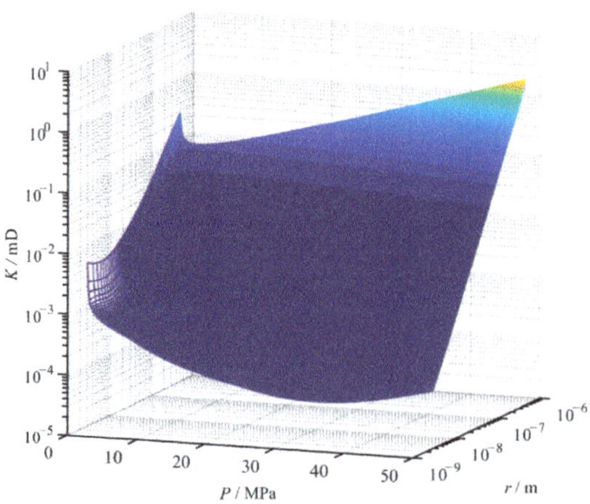

Figure 14. The relationship of the apparent permeability of shale matrix, pore size, and pressure.

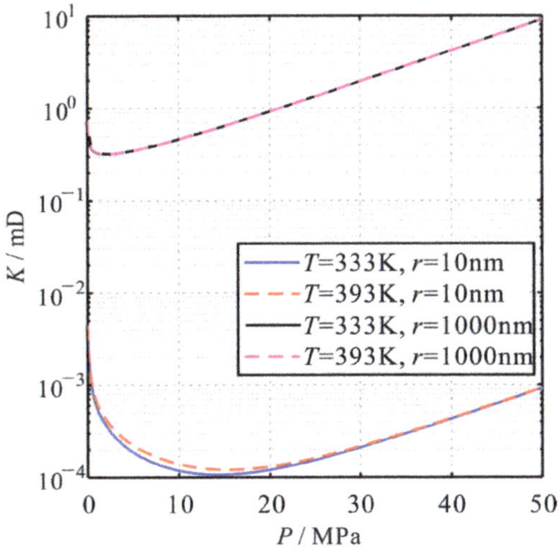

Figure 15. The relationship of the apparent permeability of shale matrix, pore size, and pressure under different temperature conditions.

The contribution ratio of each transport mechanism to the matrix's apparent permeability is further analyzed. Under the condition of constant temperature, the contribution ratio of Knudsen diffusion to the matrix's apparent permeability increased with the decrease in pressure. Moreover, its contribution ratio increased first and then decreased with the increase in pore size (see Figure 16). As Figure 17 shows, the contribution proportion of Knudsen diffusion to the apparent permeability increases with the increase in temperature, which variation decreases with the increase in pore size and pressure.

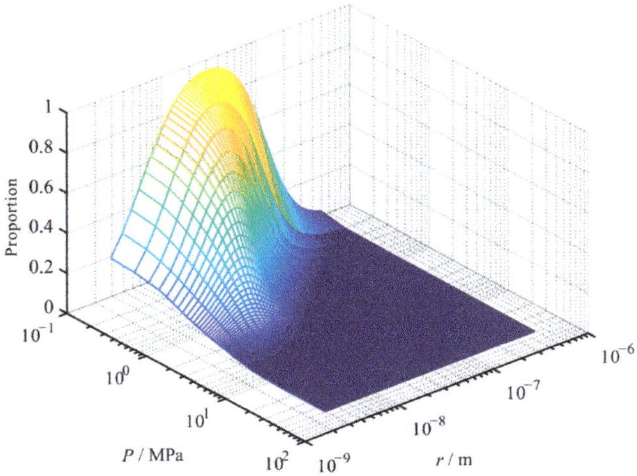

Figure 16. The various relationships for the contribution proportion of the Knudsen diffusion to the matrix apparent permeability with pore size and pressure.

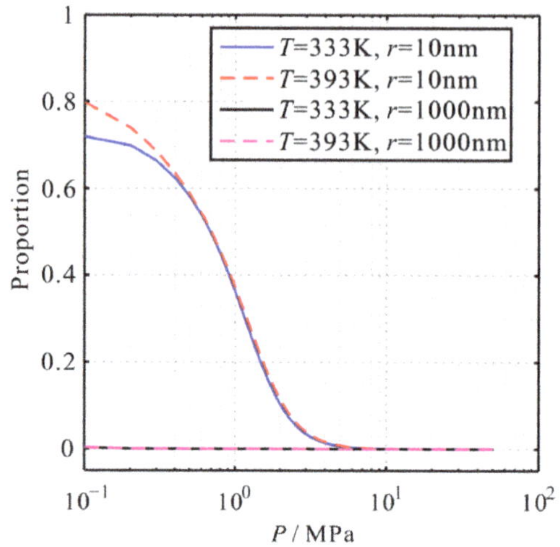

Figure 17. The various relationships for the contribution proportion of the Knudsen diffusion to the matrix apparent permeability with pore size and pressure, under different temperature conditions.

Under the condition of constant temperature, the contribution ratio of slippage flow to the matrix's apparent permeability decreased with the decrease in pressure. Moreover, its contribution ratio increased with the increase in pore size (see Figure 18). As Figure 19 shows, the contribution proportion of slippage flow to the apparent permeability decreases with the increase in temperature, which variation decreases with the increase in pore size.

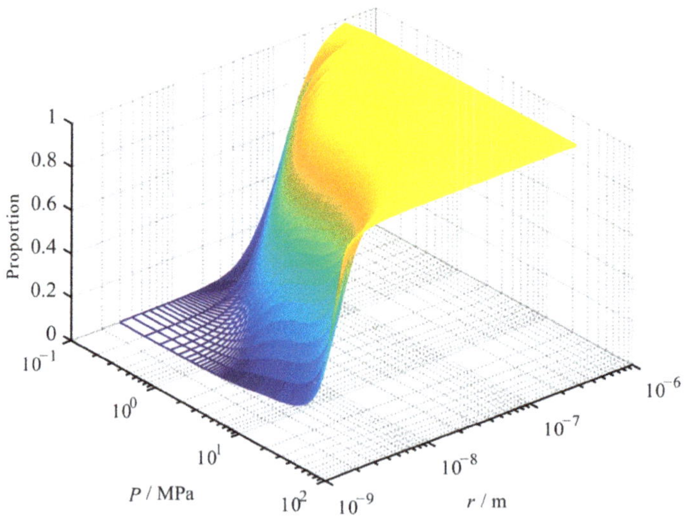

Figure 18. The various relationships for the contribution proportion of the slippage flow to the matrix apparent permeability with pore size and pressure.

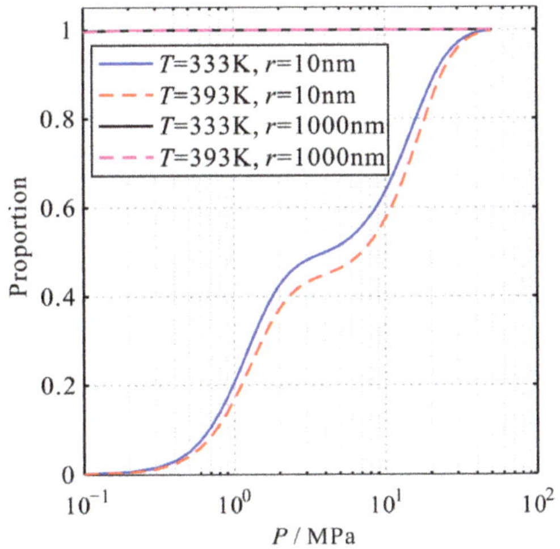

Figure 19. The various relationships for the contribution proportion of the slippage flow to the matrix apparent permeability with pore size and pressure under different temperature conditions.

Under the condition of constant temperature, the contribution ratio of surface diffusion to the matrix's apparent permeability increased first and then decreased with the increase in pressure. Moreover, its contribution ratio decreased with the increase in pore size (see Figure 20). As Figure 21 shows, the contribution proportion of surface diffusion to the apparent permeability increases with the increase in temperature, which variation decreases with the increase in pore size.

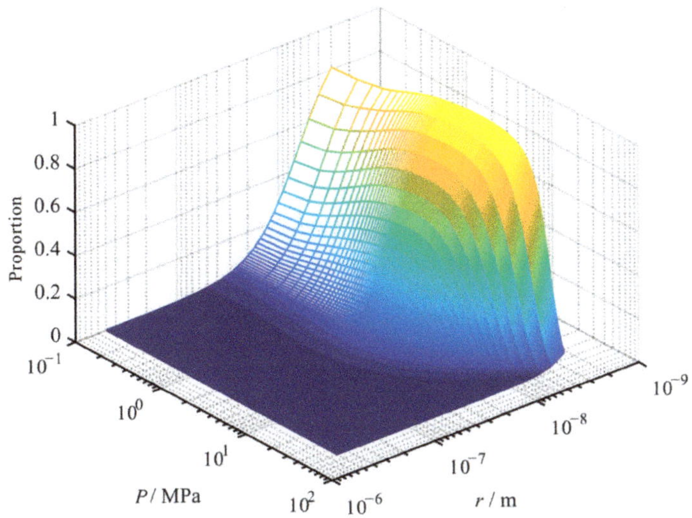

Figure 20. The various relationships for the contribution proportion of the surface diffusion to the matrix apparent permeability with pore size and pressure.

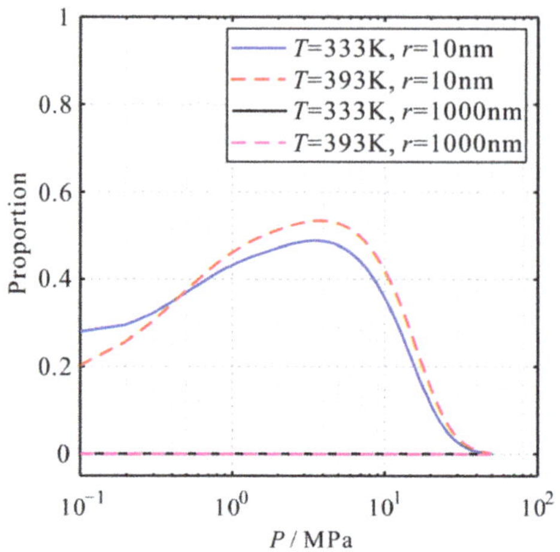

Figure 21. The various relationships for the contribution proportion of the surface diffusion to the matrix apparent permeability with pore size and pressure, under different temperature conditions.

5. Conclusions

This paper considers the changes in reservoir properties, which established a micro-transport mechanism and matrix apparent permeability model for gas in micro-nano pores and fractures of the shale matrix. The macroscopic effects under the combined action of multiple micro-transport mechanisms were studied. The conclusions are as follows:

(1) Considering the slippage flow, Knudsen diffusion, transition flow, surface diffusion, and reservoir changing mechanism, the apparent permeability model of the shale reservoir matrix is verified by the experimental data. This model can accurately fit the changes in reservoir matrix permeability in the process of shale gas development.

(2) The calculation of the Knudsen number should consider the impact of changes in reservoir and gas properties. Considering these influences, the Knudsen number decreases. Moreover, this decrease in the Knudsen number is more obvious under the conditions of low pressure and small pore size.

(3) The apparent permeability of the shale reservoir matrix is significantly influenced by pressure, temperature, pore size, and TOC. As the pressure or pore size decreases, the apparent permeability decreases first and then increases. With the increase in temperature or TOC, the apparent permeability increases. With the increase in tortuosity, the apparent permeability of the reservoir matrix decreases.

(4) The contribution proportion of the three transport mechanisms to the apparent permeability of the shale matrix is significantly influenced by pressure, temperature, pore size, and TOC. With the decrease in pressure, the proportion of slippage flow decreases, and the proportion of Knudsen diffusion increases. Moreover, the proportion of surface diffusion increases first and then decreases. With the increase in temperature, the proportion of Knudsen diffusion and surface diffusion increases, and the proportion of slippage flow decreases. With the increase in pore size, the proportion of slippage flow increases, and the proportion of surface diffusion decreases. The proportion of Knudsen diffusion increases first and then decreases. With the increase in TOC, the proportion of slippage flow decreases, and the proportion of surface diffusion increases. Moreover, the proportion of Knudsen diffusion decreases.

Author Contributions: Y.D.: developed the methodology, coded the program, and wrote the main manuscript. L.S.: discussed the methodology and conceived the work. F.L.: supervised the whole calculation. Q.Z.: discussed the model performance. C.L.: discussed the methodology. G.C.: discussed the methodology, and reviewed the manuscript. Q.C.: provided the data. S.Y.: reviewed the manuscript. J.W.: reviewed the manuscript. All authors have read and agreed to the published version of the manuscript.

Funding: This research was funded by Postdoctoral Fund Project of CNOOC Research Institute Co., Ltd., grant number 2020-KFZL-015.

Data Availability Statement: The data presented in this study are available on request from the corresponding author.

Acknowledgments: We thank all editors and anonymous reviewers for their comments and suggestions.

Conflicts of Interest: The authors declare no conflict of interest.

References

1. Wu, K.; Chen, Z. Review of gas transport in nanopores in shale gas reservoir. *Sci. Bull.* **2016**, *1*, 91–127.
2. Zhang, Q.; Liang, F.; Pang, Z.; Jiang, S.; Zhou, S.; Zhang, J. Lower threshold of pore-throat diameter for the shale gas reservoir: Experimental and molecular simulation study. *J. Pet. Sci. Eng.* **2019**, *173*, 1037–1046. [CrossRef]
3. Wang, G.; Jiang, Z.; Tang, X.; He, S.; Wang, Y.; Chang, J. Critical conditions and capabilities of shale gas diffusion and seepage types in the Longmaxi Format. *Acta Geol. Sin.* **2023**, *97*, 210–220.
4. Yang, X.; Zhou, W.; Liu, X.; Yan, Y. A multiscale approach for simulation of shale gas transport in organic nanopores. *Energy* **2020**, *210*, 118547. [CrossRef]
5. Zeng, J.; Liu, J.; Guo, J. Characterization of gas transport in shale: A multi-mechanism permeability modeling approach. *Chem. Eng. J.* **2022**, *438*, 135604. [CrossRef]
6. Wu, K.; Li, X.; Chen, Z. Micro-scale effects of gas transport in organic nanopores of shale gas reservoirs. *Nat. Gas Ind.* **2016**, *36*, 51–64.
7. Florence, F.A.; Rushing, J.; Newsham, K.E.; Blasingame, T.A. Improved permeability prediction relations for low permeability sands. In Proceedings of the Rocky Mountain Oil & Gas Technology Symposium, Denver, CO, USA, 16–18 April 2007.
8. Cipolla, C.; Lolon, E.; Mayerhofer, M. Reservoir Modeling and Production Evaluation in Shale-Gas Reservoirs. In Proceedings of the International Petroleum Technology Conference, Doha, Qatar, 7–9 December 2009. [CrossRef]
9. Sheng, M.; Li, G.; Huang, W.; Tian, S. Shale gas transient flow model with effects of surface diffusion. *Acta Pet. Sin.* **2014**, *35*, 347–352.
10. Yin, Y.; Qu, Z.; Zhang, J. An analytical model for shale gas transport in kerogen nanopores coupled with real gas effect and surface diffusion. *Fuel* **2017**, *210*, 569–577. [CrossRef]
11. Li, X.; Liu, S.; Li, J.; Tan, X. Apparent gas permeability model of shale matrix coupling stress sensitivity and water saturation. *Nat. Gas Geosci.* **2021**, *223*, 861–870.
12. Wu, K.; Li, X.; Chen, Z. Real gas transport through nanopores of shale gas reservoir. *Sci. Sin. Technol.* **2016**, *46*, 68–78. [CrossRef]
13. Ertekin, T.; King, G.A.; Schwerer, F.C. Dynamic gas slippage: A unique dual-mechanism approach to the flow of gas in tight formations. *SPE Form. Eval.* **1986**, *1*, 43–52. [CrossRef]
14. Clarkson, C.R.; Nobakht, M.; Kaviani, D.; Ertekin, T. Production analysis of tight-gas and shale-gas reservoirs using the dynamic-slippage concept. *SPE J.* **2012**, *17*, 230–242. [CrossRef]
15. Javadpour, F. Nanopores and apparent permeability of gas flow in mudrocks (shales and siltstone). *J. Can. Pet. Technol.* **2009**, *48*, 16–21. [CrossRef]
16. Swami, V.; Settari, A. A Pore Scale Gas Flow Model for Shale Gas Reservoir. In Proceedings of the American Unconventional Resources Conference, Pittsburgh, PA, USA, 5–7 June 2012.
17. Li, D.; Zhang, Y.; Sun, X.; Zhao, F.; Cui, X.; Qiao, W.; Su, Y. A new model for assessing apparent permeability of shale gas at real gas condition considering surface diffusion. *J. China Univ. Pet. (Ed. Nat. Sci.)* **2018**, *42*, 82–90.
18. Tian, S.; Wang, T.; Li, G.; Sheng, M.; Liu, Q.; Zhang, S. An analytical model for shale gas transport in circular tube pores. *Int. J. Heat Mass Transf.* **2018**, *127*, 321–328. [CrossRef]
19. Huang, T.; Tan, W.; Zhuang, Q.; Wang, G.; Yin, T. Coupling model for nanopore gas transport in shale reservoir. *J. Southwest Pet. Univ. (Sci. Technol. Ed.)* **2019**, *41*, 118–126.
20. Song, H.; Li, B.; Chen, S.; Li, J.; Gao, Z. Evolution mechanism of dynamic apparent permeability of shale reservoirs. *J. China Univ. Min. Technol.* **2022**, *248*, 873–885.
21. Villazon, G.; Sigal, R.F.; Civan, F.; Devegowda, D. Parametric Investigation of Shale Gas Production Considering Nano-Scale Pore Size Distribution. In Proceedings of the SPE Annual Technical Conference and Exhibition, Denver, CO, USA, 30 October–2 November 2011.
22. Roy, S.; Raju, R.; Chuang, H.F.; Cruden, B.A.; Meyyappan, M. Modeling gas flow through microchannels and nanopores. *J. Appl. Phys.* **2003**, *93*, 4870–4879. [CrossRef]

23. Zheng, D.; Ozbayoglu, E.; Miska, S.; Zhang, J. Experimental Study of Anisotropic Strength Properties of Shale. In Proceedings of the 57th U.S. Rock Mechanics/Geomechanics Symposium, Atlanta, GA, USA, 25–28 June 2023.
24. Zheng, D.; Ozbayoglu, E.; Miska, S.; Zhang, J. Combined Experimental and Well Log Study of Anisotropic Strength of Shale. In Proceedings of the SPE Annual Technical Conference and Exhibition, San Antonio, TX, USA, 16–18 October 2023.
25. Liu, J.; Bai, X.; Elsworth, D. Evolution of pore systems in low-maturity oil shales during thermal upgrading—Quantified by dynamic SEM and machine learning. *Pet. Sci.* **2024**, *in press*. [CrossRef]
26. Javadpour, F.; Fisher, D.; Unsworth, M. Nanoscale gas flow in shale gas sediments. *J. Can. Pet. Technol.* **2007**, *46*, 55–61. [CrossRef]
27. Wu, K.; Chen, Z.; Li, X.; Guo, C.; Wei, M. A model for multiple transport mechanisms through nanopores of shale gas reservoirs with real gas effect-adsorption-mechanic coupling. *Int. J. Heat Mass Transf.* **2016**, *93*, 408–426. [CrossRef]
28. Shi, J.; Lei, Z.; Li, Y.; Wei, Y.; Tao, W. Diffusion and flow mechanisms of shale gas through matrix pores and gas production forecasting. In Proceedings of the SPE Unconventional Reservoir Conference, Calgary, AB, Canada, 5–7 November 2013.
29. Raghavan, R.; Chin, L.Y. Productivity changes in reservoirs with stress-dependent permeability. *SPE Reserv. Eval. Eng.* **2004**, *7*, 308–315. [CrossRef]
30. Fu, X.; Li, D.; Qin, Y.; Jiang, B.; Wang, W.; Li, G. Experimental Research of Influence of Coal Matrix Shrinkage on Permeability. *J. China Univ. Min. Technol.* **2002**, *31*, 129–131.
31. Cao, C.; Li, T.; Zhang, L.; Gao, C.; Wang, H. Shale gas dual porosity-dual permeability model with matrix shrinking. *Nat. Gas Geosci.* **2015**, *26*, 2381–2387.
32. Wang, S.; Elsworth, D.; Liu, J. A mechanistic model for permeability evolution in fractured sorbing media. *J. Geophys. Res.* **2012**, *117*, 6205–6221. [CrossRef]
33. Dong, Y.; Zhao, Q.; Lu, C.; Lu, C.; Gan, Y.; Chen, G. High-pressure isothermal adsorption model for supercritical methane in shale. In Proceedings of the 2022 International Field Exploration and Development Conference, IFEDC, Xi'an, China, 16–18 November 2022; Springer Science and Business Media Deutschland GmbH: Singapore, 2022.
34. Shi, R.; Liu, J.; Derek, E. Mechanistic Analysis of Shale Permeability Evolution Data. In Proceedings of the Unconventional Resources Technology Conference, Denver, CO, USA, 22–24 July 2019.

energies

MDPI

Article

Comparison of the Sample Preparation Strategies and Impacts on the Tensile Strength of Gas Shale with Variable Moisture Conditions

Liuqing Shu [1], Lingzhi Xie [1,2], Bo He [2,3,*] and Yao Zhang [1]

[1] College of Architecture & Environment, Sichuan University, Chengdu 610065, China; scu_shuliuqing@163.com (L.S.); xielingzhi@scu.edu.cn (L.X.); zhangyao@scu.edu.cn (Y.Z.)
[2] Institute of New Energy and Low-Carbon Technology, Sichuan University, Chengdu 610065, China
[3] Engineering Research Center of Alternative Energy Materials and Devices, Ministry of Education, Sichuan University, Chengdu 610065, China
* Correspondence: hebo@scu.edu.cn; Tel.: +86-159-8205-2378

Abstract: Moisture significantly affects the mechanical behavior of gas shale and further determines the hydraulic fracturing performance, as it is more attractive. Nevertheless, batch experiments have usually involved variable methodologies regarding the preparation of moisture-contained shale specimens in the sequence (and/or frequency) of drying and soaking treatments. Accordingly, this work investigates how the preparation methodology influences the test results of moisture-contained shale samples. This study compares three commonly used shale sample preparation strategies for acquiring different moisture contents, that is, "dry-wet", "dry-wet-dry", and "wet-dry-wet" strategies, followed by a Brazilian splitting test for the mechanical parameters. The results show that under the same saturation conditions, the longer the soaking time during sample preparation, the higher the degradation degree of shale tensile strength. Meanwhile, prolonged soaking can lead to a more discrete distribution of strength values, and the failure mode may deviate from the Brazilian splitting theory model. Under the combined influence of moisture content and soaking time, the tensile strength of shale decreases approximately linearly with increasing saturation, while the degradation degree increases nonlinearly with increasing saturation, and the degradation rate changes from slow to fast. According to the observation of the microstructure of hydrated shale, prolonged soaking can lead to an increase in the expansion of clay minerals in shale by hydration, resulting in looser and more fragmented internal structure, and further degradation in shale strength. In order to weaken the interference of hydration when studying the effect of moisture content on the tensile strength of shale, the soaking time should be minimized as much as possible during the preparation process.

Keywords: shale mechanics; moisture saturation; tensile strength; Brazilian splitting; soaking time

1. Introduction

For a long time, coal has been the mainstay of China's energy structure. With the increase in energy demand, the government actively promotes the optimization and transformation of the energy structure, which drives the development of unconventional geological energy suppliers, like shale gas, shale oil, and gas hydrate [1–3]. At the beginning of the 21st century, the United States completed the shale gas revolution and realized the strategic goal of energy independence, which led the world to thoroughly consider the possibility of shale gas exploitation [4]. According to the assessment of the U.S. Energy Information Administration (E.I.A.), China has abundant shale gas resources with the technically recoverable amount of shale gas resources as high as 31.57×10^{12} m^3 which ranks first in the world [5]. Shale gas will become a new field for the development of green energy in China due to its considerable resource potential and cleanliness advantages [6].

It is necessary to stimulate the reservoir through hydraulic fracturing technology to increase shale gas production because of the low porosity and permeability of shale, and the tensile strength of shale is also an important parameter in this process [7–11]. However, substantial dispersions in the moisture content of shale reservoirs are evident across different regions [12]. Li et al. [13] found that the moisture content can affect the tensile strength of shale, and the water will cause strength degradation. Zheng et al. [14,15] found that shale bedding has a significant impact on the hydraulic fracturing effect when studying the anisotropic characteristics of shale, and provided innovative ideas for solving the instability problem during the fracturing process by combining experimental results with logging data. Teng et al. [16] considered the bedding properties of shale when discussing the effect of different water contents on the tensile strength of shale, and found that the damage of water to shale strength is mainly achieved through bedding.

The systematic study of the influence of moisture content on the tensile strength of shale is of great significance for the hydraulic fracturing effect. However, the preparation strategies for water-bearing rocks are diverse, there is no unified standard, and the differences in preparation strategies are mainly reflected in the order and frequency of soaking and drying treatments. Debanjan et al. [17] dried shale and sandstone for 24 h, saturated them with distilled water, and tested their degradation at different saturation levels via Brazilian splitting. Debanjan et al. found that shale degraded quickly while sandstone degraded slower until 80% saturation. Babets et al. [18] saturated samples first, then dried them, finding that the strength of siltstones and mudstones decreased 1.5–2.5 times in a water-saturated state compared to when air-dried. Hasbollah et al. [19] dried and partially soaked shale and sandstone, observing that the tensile strength and anisotropy rose with water content, and more so in shale. Lin et al. [20] obtained approximate saturation by estimating shale porosity and pore volume, conducting adsorption tests that showed gas adsorption capacity fell with increasing water saturation. Discrepancies in sample preparation strategies may influence the outcomes of experiments when exploring the influence of moisture content on the mechanical properties of rocks, thereby resulting in insufficient grounds for comparing and correlating different research findings [21].

Presently, discussions on the deterioration of shale strength due to moisture content encounter challenges such as varying sampling positions and the absence of standardized sample preparation strategies, and there is a lack of research on the influence of moisture content on the strength of the same rock under different preparation strategies. Furthermore, the disparate laws and extents of shale tensile strength deterioration elucidated by different research outcomes indicate a lack of comparative studies on different sample preparation protocols under identical geological conditions. Therefore, this paper devises three distinct shale sample preparation protocols for Longmaxi shale, conducts Brazilian splitting tests under different saturation conditions, compares the impacts of sample preparation protocols on the dispersion of shale tensile strength under saturated conditions and their failure modes, and analyzes the mechanism of this influence from a microstructural perspective. Finally, the test results of the three sample preparation protocols are scrutinized to offer insights for selecting sample preparation protocols for different moisture contents.

2. Materials and Methods

2.1. Preparation Strategies of Samples with Different Saturation

We designed different sample preparation strategies based on the sequence and frequency of drying/wetting treatments and named them accordingly. The specific strategies are as follows:

For the convenience of comparing the test results of different preparation strategies, saturation is considered a reasonable characterization parameter which is obtained by the weighing method. According to the ISRM recommended method [22], the saturation calculation formula is:

$$s = \frac{m_0 - m_{\text{dry}}}{m_{\text{sat}} - m_{\text{dry}}} \times 100\% \quad (1)$$

where s is the saturation; m_0 is the mass of the sample to be tested (g); m_{dry} is the mass of the sample when it is completely dry (g); and m_{sat} is the mass when it is completely saturated (g).

The samples were immersed in water at room temperature until fully saturated (the mass was recorded every two hours, and when the mass remained constant for three consecutive measurements, it was considered saturated, a process that took approximately 24 h). This process was considered a complete "wet treatment". Since this paper focuses on analyzing the effects of water on sample preparation, a lower temperature was used to dry the samples to avoid thermal damage and changes in pore distribution caused by high temperature in the sample and then affect the mechanical properties and microstructure of rocks [23–26]. The samples were dried at 55 °C until completely dry (the mass remained unchanged for three consecutive measurements after approximately 24 h of drying), considered a complete "dry treatment".

Based on the sequence and frequency differences between a "dry treatment" and "wet treatment," three sample preparation strategies were designed. The schematic diagrams of each preparation process are shown in Figure 1.

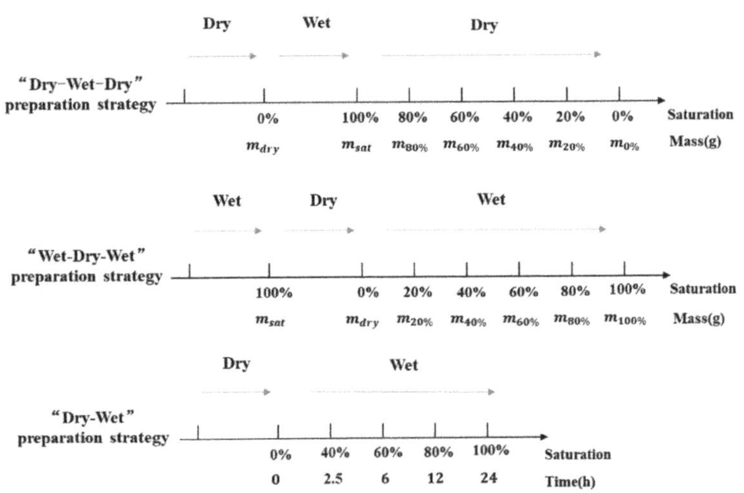

Figure 1. Schematic diagram of the three sample preparation strategies.

In the "dry-wet-dry" preparation strategy, the samples were placed in a drying oven at 55 °C for 24 h to achieve complete drying, obtaining the sample mass (m_{dry}). Subsequently, the samples were soaked in water for 24 h to achieve complete saturation, obtaining the sample mass (m_{sat}). Finally, the samples were subjected to further drying at 55 °C until the sample mass approached the theoretical mass corresponding to target saturations of 0%, 20%, 40%, 60%, 80%, and 100% (according to Equation (1)).

In the "wet-dry-wet" preparation strategy, the samples were soaked in water for 24 h to achieve complete saturation, obtaining the sample mass (m_{sat}). Then, the samples were placed in a drying oven at 55 °C for 24 h to achieve complete drying, obtaining the sample mass (m_{dry}). Finally, the samples were subjected to further soaking until the sample mass approached the theoretical mass corresponding to target saturations of 100%, 80%, 60%, 40%, 20%, and 0% (according to Equation (1)).

In the "dry-wet" preparation strategy, the samples were placed in a drying oven at 55 °C for 24 h to achieve complete drying, obtaining the sample mass (m_{dry}). Then, the samples were soaked in water for 2.5 h, 6 h, 12 h, and 24 h to obtain samples with target saturations of 0%, 40%, 60%, 80%, and 100%, and the soaking time corresponding to the saturation is obtained by referring to the pre-immersion test. Due to the inability to ensure each sample reach complete saturation under this strategy, it is not possible to accurately obtain the corresponding mass of the sample when reaching the target saturation through calculation like the previous two strategies. Therefore, only an approximate saturation can be obtained by controlling the soaking time.

The number of prepared samples and completed loading tests is summarized in Table 1. The samples fail before loading during the soaking time because of the hydration cracks, which cause a decrease in the availability of test results for some sample groups.

Table 1. Summary of number of prepared samples.

Load Mode	Target Saturation	Preparation Strategies		
		"Dry-Wet-Dry"	"Wet-Dry-Wet"	"Dry-Wet"
Perpendicular to bedding	0%	3	3	5
	20%	2	3	/
	40%	1	4	4
	60%	3	5	2
	80%	3	3	3
	100%	4	3	3
Parallel to bedding	0%	2	4	6
	20%	2	4	/
	40%	4	6	3
	60%	4	4	1
	80%	3	5	3
	100%	3	4	3

2.2. Test Methods

The sample material was collected from the Longmaxi Formation shale outcrop in Lu Jiao Town, Pengshui Miao, and Tujia Autonomous County, Chongqing City. The main minerals are quartz, sodium feldspar, and muscovite, with relative contents of 46%, 31%, and 23%, respectively [27]. The shale collected is gray-black, with distinct bedding structures that are easily discernible.

Irregular body outcrops were cut into rectangular prisms with bedding, and rock cores with a diameter of 50 mm were drilled using a dry drilling strategy. The rock cores were then cut into circular disc samples with dimensions of Φ50 mm $\times$ H25 mm, and their surfaces were polished. The parallelism of the upper and lower surfaces of the disc samples was controlled within 0.5 mm, and the surface roughness was controlled within 0.1 mm.

The processed samples were treated with different sample preparation strategies to obtain different levels of saturation. Brazilian splitting tests were conducted on samples with different saturations using a rock mechanic testing machine (maximum load 10 kN, measuring accuracy 1 N). The test design included two loading modes: parallel bedding and perpendicular bedding. The loading process used axial displacement as the control index and loaded the sample at a stable rate of 0.05 mm/min until failure. The preparation and loading of the sample are shown in Figure 2.

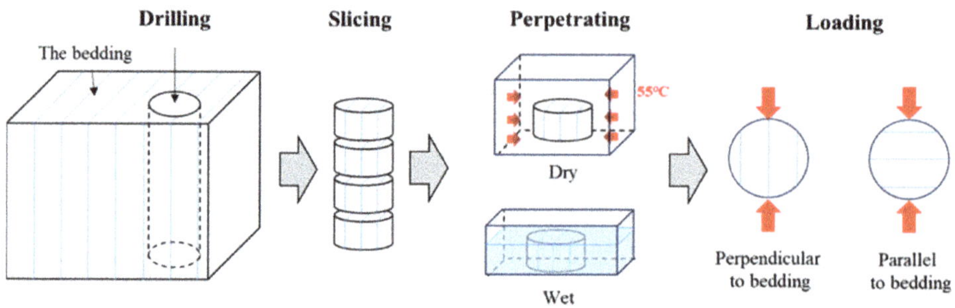

Figure 2. Sample processing, preparation, and loading.

The formula for calculating the tensile strength of samples using the Brazilian splitting test is as follows [28]:

$$\sigma_t = \frac{2F}{\pi d h} \tag{2}$$

where σ_t is the tensile strength of the sample (MPa); F is the load when the sample is destroyed (N); d is the diameter of the sample (mm); and h is the thickness of the sample (mm).

3. Results

3.1. Relationship between Shale Tensile Strength and Saturation

The tensile strength–saturation curves of the samples prepared using three different strategies under parallel and perpendicular bedding loading modes are shown in Figure 3.

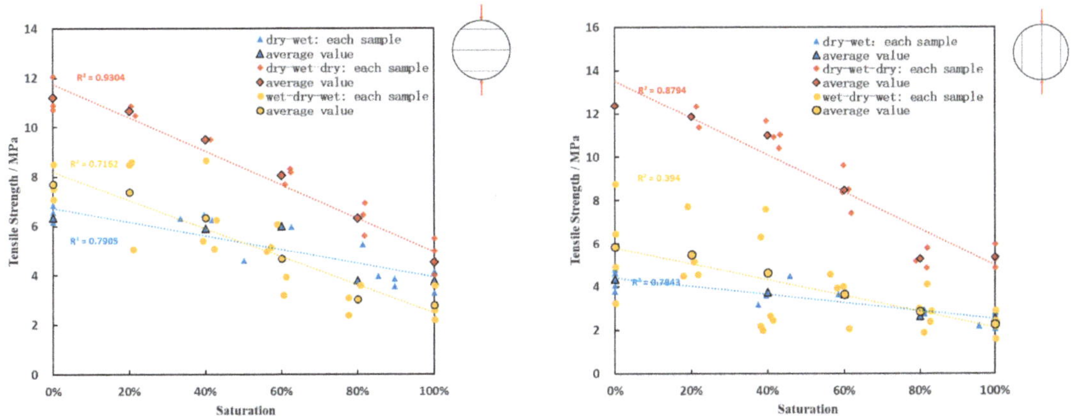

(a) (b)

Figure 3. Tensile strength–saturation: (**a**) perpendicular bedding loading; (**b**) parallel bedding loading.

From Figure 3, it can be observed that the tensile strength of shale decreases significantly with increasing saturation, showing an approximate linear change trend, which is independent of the selection of sample preparation strategies and loading directions. In the "dry-wet" preparation group, the tensile strength of shale is 3.81–6.84 MPa in the completely dry state and 2.06–4.16 MPa in the completely saturated state, with strength differences of 2.72 MPa and 1.81 MPa under perpendicular and parallel bedding loading, respectively. In the "dry-wet-dry" preparation group, the tensile strength of shale is 10.71–12.40 MPa in the completely dry state and 3.61–5.95 MPa in the completely saturated state, with strength differences of 6.69 MPa and 7.04 MPa under perpendicular and parallel bedding loading, respectively. In the "wet-dry-wet" preparation group, the tensile strength of shale is 3.26–8.52 MPa in the completely dry state and 1.60–3.60 MPa in the completely saturated state, with strength differences of 4.90 MPa and 3.58 MPa under perpendicular and parallel bedding loading, respectively.

3.2. Tensile Strength Dispersion

To analyze the influence of different sample preparation strategies on the tensile strength dispersion of saturated shale, the formula of the coefficient of dispersion is as follows [29]:

$$V_s = \frac{\sigma}{\mu} \tag{3}$$

where V_s is the coefficient of dispersion; σ is the standard deviation of shale tensile strength (MPa); and μ is the average value of shale tensile strength (MPa).

The tensile strength dispersion of shale samples under different saturation for three sample preparation strategies is shown in Figure 4. The coefficient of dispersion of shale tensile strength in the "dry-wet" preparation group ranges from 0 to 0.1460, in the "dry-wet-dry" preparation group it ranges from 0 to 0.1668, and in the "wet-dry-wet" preparation group it ranges from 0.0784 to 0.5068, with most values exceeding 0.2. When the coefficient of dispersion is 0, the number of samples is 1.0 and is not discussed. Under the same saturation conditions, the coefficient of dispersion of tensile strength is relatively small for the "dry-wet" preparation group and "dry-wet-dry" preparation group, while it is consistently higher for the "wet-dry-wet" preparation group, indicating that the differences in sample preparation strategies affect the tensile strength dispersion of saturated shale.

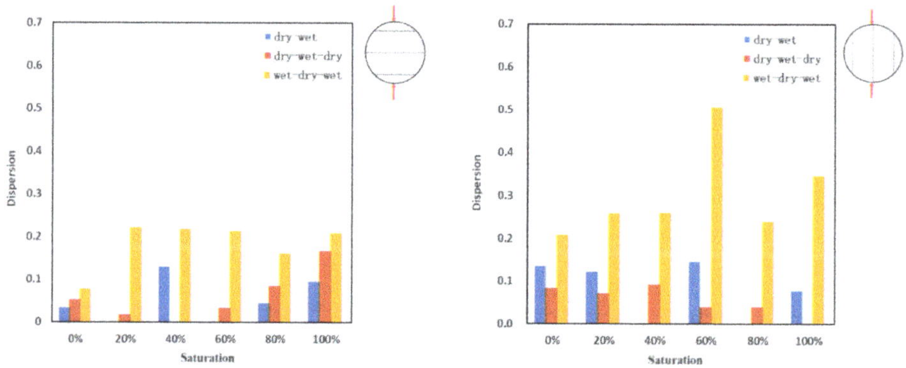

(**a**) (**b**)

Figure 4. Dispersion–saturation: (**a**) perpendicular bedding loading; (**b**) parallel bedding loading.

3.3. *Relationship between Tensile Strength Degradation and Saturation*

In the process of studying the impact of fluids on rock strength, it is found that various types of rocks exhibit a law of strength degradation due to their moisture content, and the degree of degradation is quantified using the degradation degree, and the formula is the following [15]:

$$D_{\mathrm{B.T.S.}} = \left(1 - \frac{\sigma_{\mathrm{t0}}}{\sigma_{\mathrm{tdry}}}\right) \times 100\% \tag{4}$$

where $D_{\mathrm{B.T.S.}}$ is the degree of Brazilian tensile strength degradation; σ_{t0} is the tensile strength of the sample at any saturation (MPa); and σ_{tdry} is the tensile strength of the sample when completely dry (MPa).

The relationship between the degree of degradation and the saturation of shale samples prepared using three sample preparation strategies is shown in Figure 5. With increasing saturation, the degree of degradation of the sample also increases, and the degree of the degradation–saturation curve exhibits an upward concave shape, indicating that the strength degradation of samples becomes more severe with increasing saturation, particularly at high saturation. Although changes in sample preparation strategies and loading directions do not affect the law of degradation degree dispersion with saturation, the choice of sample preparation strategies leads to differences in the degree of degradation of samples under different moisture contents.

When the sample saturation is low, the degradation degrees of the samples under the "dry-wet-dry" and "wet-dry-wet" strategies are similar, while the degradation degree of the sample under the "dry-wet" strategy is significantly smaller, and the difference between the strategies is more obvious with the gradual increase in saturation. During the test, when the saturation was 0%, the "dry-wet-dry" and "wet-dry-wet" samples had undergone 24 h wet treatment, so the degradation degree was similar, while the "dry-wet" samples had not been wet treated, but the degradation degree was lower than the other two. With the increase in saturation, the contact time of samples with different strategies with water increases.

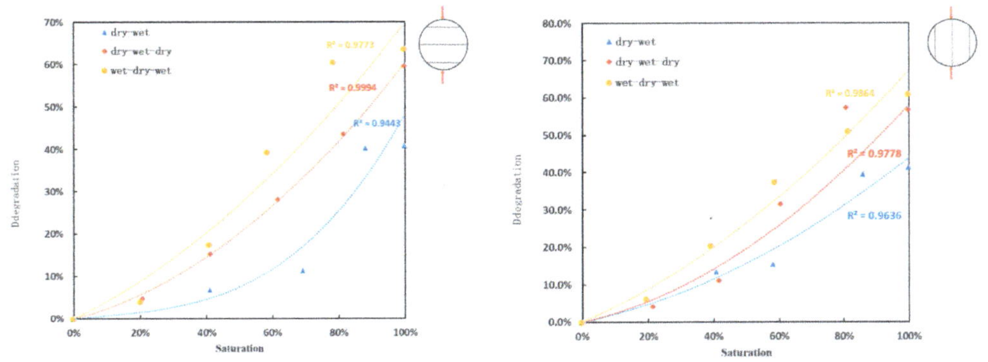

(**a**) (**b**)

Figure 5. Degradation degree–saturation: (**a**) perpendicular bedding loading; (**b**) parallel bedding loading.

Comparing the degradation degree of shale samples prepared using three sample preparation strategies at different saturations, the results are shown in Table 2. With increasing saturation, the degradation degree of shale samples prepared using the "dry-wet" preparation strategy ultimately increases to about 40%, while those prepared using the "wet-dry-wet" and "dry-wet-dry" preparation strategies increase to about 60%. It is worth noting that in the "wet-dry-wet" preparation process, the sample fails before loading when fully saturated, indicating that the strength of the sample is still deteriorating, and the actual degradation may be higher than the test results. Overall, the deteriorative degree of samples prepared using the "wet-dry-wet" strategy is relatively higher among the three. In contrast, the deteriorative degree of samples prepared using the "dry-wet" strategy is relatively lower, and the deteriorative degree of samples prepared using the "dry-wet-dry" strategy is between the two.

Table 2. Results of degradation degree.

Saturation	Perpendicular Bedding Loading			Parallel Bedding Loading		
	"Dry-Wet"	"Dry-Wet-Dry"	"Wet-Dry-Wet"	"Dry-Wet"	"Dry-Wet-Dry"	"Wet-Dry-Wet"
100%	40.8%	59.6%	63.6%	41.4%	57.4%	61.1%
80%	40.3%	43.6%	60.5%	39.5%	31.6%	51.2%
60%	11.5%	28.2%	39.4%	15.6%	11.1%	37.5%
40%	6.9%	15.3%	17.5%	13.6%	4.3%	20.5%
20%		4.9%	4.2%		4.9%	6.3%

3.4. Failure Mode

Yang et al. [30] categorized the Brazilian splitting failure modes of shale into crescent, arc, and straight-line types based on the angle between the bedding and loading direction. The shale samples exhibit a straight-line failure mode when the loading direction is perpendicular or parallel to the bedding.

The failure modes of samples prepared using the three different sample preparation strategies in completely dry and completely saturated states are shown in Figure 6. The results indicate that the choice of sample preparation strategy affects the Brazilian splitting failure mode of water-bearing shale.

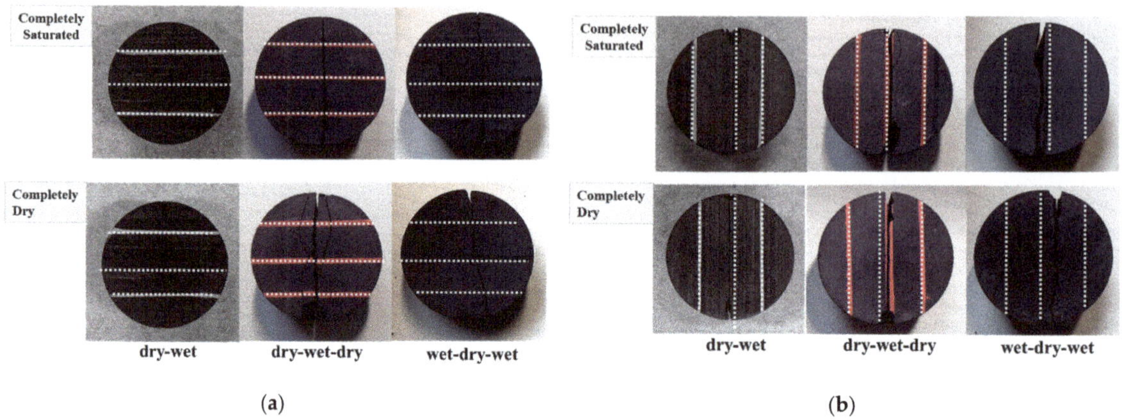

Figure 6. Destruction mode: (**a**) perpendicular bedding loading; (**b**) parallel bedding loading.

The bedding structure of shale is symmetric about the loading direction in both parallel and perpendicular bedding loading modes, resulting in a symmetrical stress field about the loading surface, leading to straight-line cracks through the center of the sample during shale failure. When the "dry-wet" and "dry-wet-dry" preparation strategies are chosen, the main crack in the shale splits along the sample's central axis under completely dry and completely saturated conditions, forming relatively smooth straight lines. When the "wet-dry-wet" preparation strategy is chosen, the shale sample still exhibits a straight-line failure mode when completely dry. However, under completely saturated conditions, the crack path of the sample deviates from the center position, and the crack morphology is no longer smooth but shows multiple inflection points, presenting a non-linear failure pattern. These results indicate that the internal structure of shale changes significantly under this preparation condition, affecting the symmetry of the stress field distribution, and the complex stress distribution leads to the manifestation of shear failure characteristics in shale. The requirement for a more extended sample soaking time in the "wet-dry-wet" preparation process to achieve complete saturation and mechanical testing is a significant reason for the non-linear failure mode of the sample. The characteristics of the two types of failure modes are shown in Figure 7.

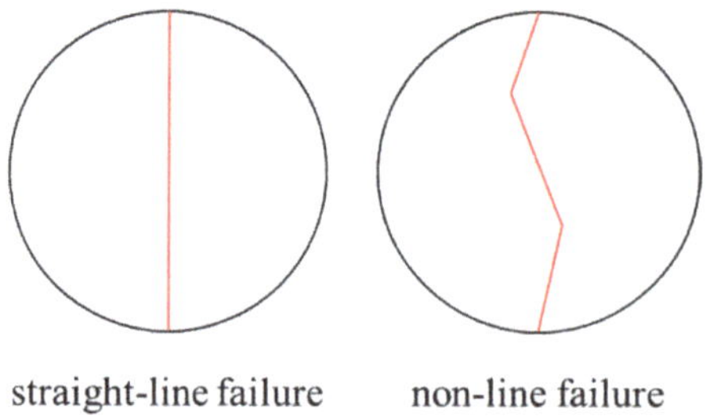

Figure 7. Two characteristic failure modes.

3.5. Mechanism Analysis

Shale is primarily composed of clay minerals such as illite, montmorillonite, kaolinite, and chlorite, as well as quartz, muscovite, pyrite, and other minerals. The hydration expansion characteristics of clay minerals significantly impact the mechanical properties of shale when exposed to water [31]. The capillary effect, as a crucial prerequisite for shale hydration, promotes water penetration into the rock's interior and contact with clay minerals, initiating hydration reactions during the sample soaking process [32]. Clay mineral hydration reactions mainly involve two stages: surface hydration and osmotic hydration. Surface hydration occurs as water molecules form several molecular layers on the clay mineral surface through adsorption, causing mineral lattice expansion. Osmotic hydration manifests as a concentration gradient between the high-charge cations of small radius in the interlayer of mineral layers and the cations in the soaking water, leading to water penetration between the layers and increasing the interlayer spacing. The changes in lattice structure and interlayer spacing expansion result in the volumetric expansion of clay mineral particles. Observations of hydrated shale microstructures using techniques such as CT imaging and electron microscopy reveal that clay mineral expansion leads to loosening of the internal structure, fragmentation, and an increase in micropores and microcracks [23,33,34]. These changes deepen the influence of capillary forces on the internal shale cracks and, combined with the hydration stress generated by clay mineral expansion, contribute to crack propagation [35]. The structural changes in shale induced by capillary forces and clay mineral hydration are irreversible and tend to become looser and more fragmented over time due to hydration effects, ultimately leading to a further reduction in shale strength. Hydration in shale begins during water absorption, with noticeable hydration cracks observed within 24 h of immersion [36,37].

Moreover, the expansion degree of clay mineral hydration increases with soaking time, indicating a significant temporal dependency in the deterioration in hydrated shale strength [38]. However, the low-temperature drying of shale at 55 °C does not have a significant impact, and it is also insufficient to completely remove the adsorbed water and interlayer water on clay mineral particle surfaces (the temperature needs to reach 110 °C) [39]. Therefore, dry treatment does not affect the interaction between clay particles and water in shale but only influences the presence of free water in shale. During the preparation of hydrated shale specimens, capillary effects and clay mineral hydration expansion cause internal structure loosening, fracturing, crack development, and consequently, a decrease in shale strength, with the degree of influence depending on the soaking time in different preparation strategies.

In the "wet-dry-wet" sample preparation strategy, SEM scanning was performed when the specimens reached complete saturation after the first and second wet treatments, producing images as shown in Figure 8. Figure 8a shows that even when the specimen reaches complete saturation after the first wet treatment, tiny gaps still exist between the minerals inside the shale, and the mineral surfaces exhibit granular materials. Figure 8b reveals that after complete drying and subsequent re-saturation during the second wet treatment, clay minerals exhibit significant expansion. It can be seen from Figure 8c that the minerals are exfoliated after the shale is soaked in water, and the small broken particles are distributed on the surface of the sample. After the second wet treatment, the number of these broken particles increases, as shown in Figure 8d. However, it can be seen in Figure 8e that after the first wet treatment, there are a small number of tiny pores between minerals (pyrite in the figure). After secondary immersion, these micropores tend to increase, as shown in Figure 8f. This indicates that with increased wet treatment time, the internal structure of shale, even under complete saturation, becomes looser and more fragmented. Furthermore, shale hydration exhibits time-dependent characteristics, and prolonged immersion can deepen hydration, leading to further deterioration in shale tensile strength.

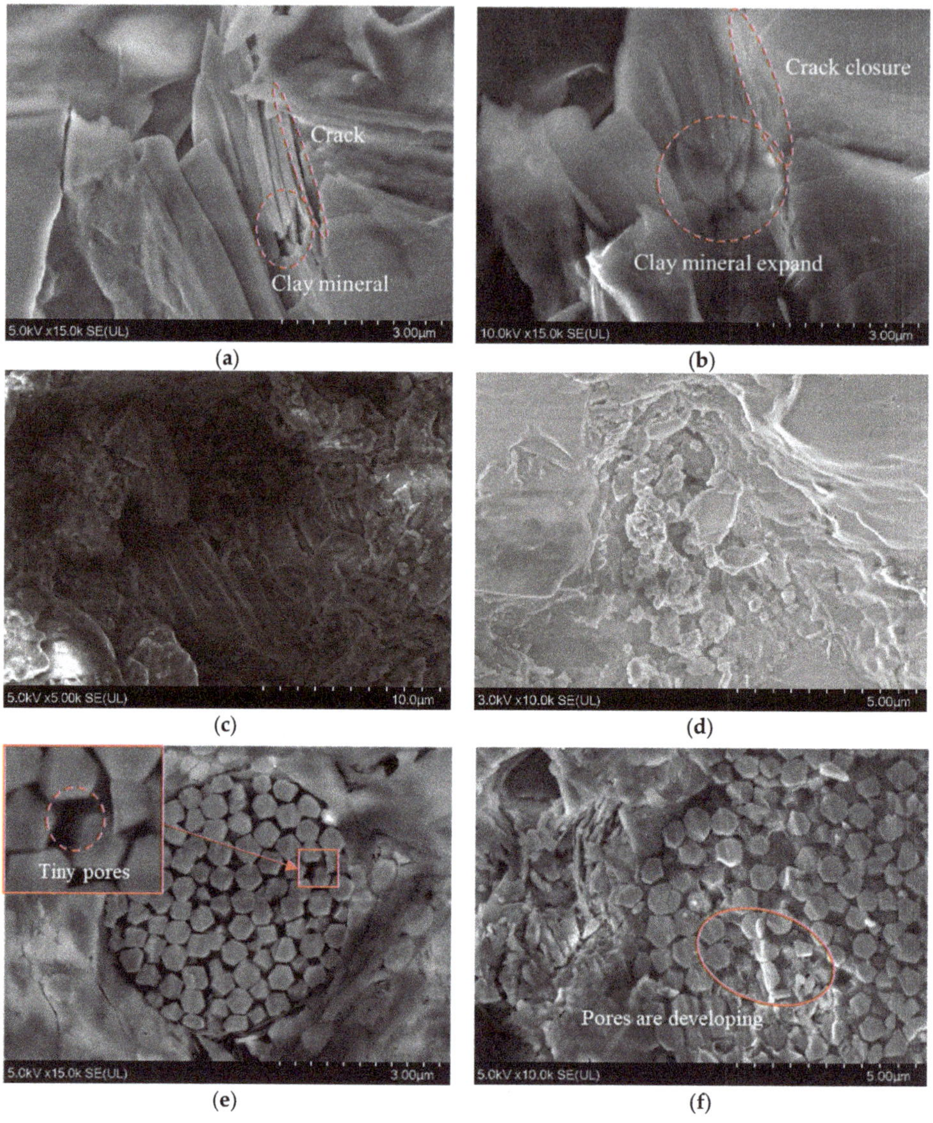

(a) (b) (c) (d) (e) (f)

Figure 8. SEM. images of shale during the "wet-dry-wet" sample preparation process: (**a**,**c**,**e**) the first wet treatment reaches saturation; (**b**,**d**,**f**) the second wet treatment reaches saturation.

The differences in the three sample preparation strategies mainly stem from the number of wet treatment times and their durations. The wet treatment time required to achieve complete saturation of the sample is approximately 24 h, while the wet treatment time for preparing samples with different saturation ranges from 0 to 24 h. Thus, both the "dry-wet" and "dry-wet-dry" sample preparation strategies involve one wet treatment, with wet treatment times of 0–24 h and 24 h, respectively, while the "wet-dry-wet" strategy involves two wet treatments with wet treatment times of 24–48 h. During sample preparation, the wet treatment processes lead to shale hydration, and longer wet treatment times result in increased shale hydration, leading to further deterioration in shale strength, increased dispersion in test results, and deviation from theoretical models in sample failure modes.

In summary, the increase in wet treatment time and frequency during sample preparation strengthens the changes in shale internal structure induced by capillary forces and clay mineral hydration expansion, resulting in increased degradation and dispersion in shale tensile strength, and influencing the failure mode of the shale specimens.

4. Discussion

Since shale water absorption saturation and hydration reactions occur simultaneously during sample preparation [35], under this experimental condition, the deterioration in shale strength represents the combined result of wet treatment time and moisture content, making it difficult to isolate the individual effects of saturation or moisture content on shale tensile strength using conventional sample preparation strategies. Moreover, moisture content alone cannot reflect the degree of hydration, and explaining the mechanism of shale strength degradation caused by moisture content from a hydration perspective lacks sufficient evidence. Therefore, only approximate conclusions can be drawn regarding the effects of wet treatment time on shale strength deterioration.

Wet treatment time is a crucial factor affecting the shale hydration reaction, and during the preparation of samples with different moisture contents, the wet treatment time is influenced by the selection of sample preparation strategies. Even when samples with the same saturation or moisture content are prepared using different methods, the degree of degradation in shale strength varies. Strategies with longer wet treatment times intensify shale hydration, which causes further deterioration in shale strength, dispersion in the test results, and deviation from theoretical models in sample failure modes. The problem of increased tensile strength dispersion caused by shale hydration also hampers the determination of the relationship between tensile strength and saturation. When the sample exhibits a non-linear failure pattern, the stress characteristics deviate from the theoretical model of the Brazilian splitting test, making it more challenging to characterize the actual strength of the sample using the theoretical model.

From the perspective of tensile strength dispersion, both the "dry-wet" and "dry-wet-dry" sample preparation strategies yield results below 0.17, while the "wet-dry-wet" strategy yields results mostly exceeding 0.2, indicating significantly greater dispersion. Analyzing the degree of tensile strength degradation, the maximum degradation in the "dry-wet" strategy reaches 40%. In contrast, in the "dry-wet-dry" and "wet-dry-wet" strategies, the maximum degradation can be 60%, with the "wet-dry-wet" shale exhibiting immediate degradation before loading, suggesting potentially more significant degradation. Thus, the "dry-wet" sample preparation strategy effectively reduces the degradation in sample tensile strength. Regarding failure modes, both the "dry-wet" and "dry-wet-dry" strategies exhibit linear failure patterns during the transition from complete drying to complete saturation, consistent with the theoretical model of Brazilian splitting. However, after "wet-dry-wet" sample preparation, samples under complete saturation exhibit a non-linear failure pattern, indicative of shear failure characteristics, deviating from the theoretical model.

To better discuss the influence of moisture content on shale tensile strength and minimize unfavorable factors in the test results' analysis, efforts should be made to mitigate the degradation in shale strength caused by clay mineral hydration. In selecting sample preparation strategies, this goal can be achieved by reducing the wet treatment time as much as possible. The "dry-wet" sample preparation strategy effectively shortens the wet treatment time, allowing test results to reflect the impact of moisture content on shale tensile strength with increased accuracy and facilitating a systematic summary of the results.

Additionally, other strategies can be employed to mitigate the influence of shale hydration reactions on research results. For instance, solutions that inhibit shale hydration can be used as soaking liquids [40]. Furthermore, wet treatment strategies and equipment improvements can further shorten the wet treatment times [41]. However, these strategies only reduce the impact of shale hydration and cannot eliminate it. Therefore, when studying the effects of different moisture contents on shale tensile strength deterioration, the influence of shale hydration will always be present, and test results can only approximate the relationship between moisture content and shale tensile strength.

5. Conclusions

This study investigated the influence of sample preparation strategies on the variation in shale tensile strength with moisture content, employing three different saturation sample preparation strategies in Brazilian splitting tests. The following conclusions were drawn:

(1) The soaking time when preparing the sample will affect the degradation effect of shale strength, that is, the longer the soaking time is, the higher is the strength degradation degree of the sample with the same moisture content. Therefore, in the process of studying the effect of different moisture contents on the tensile strength of shale, the effect of soaking time cannot be ignored.
(2) Under the joint influence of moisture content and soaking time, the tensile strength of shale decreases approximately linearly with the increase in saturation. The degree of deterioration increases nonlinearly with the increase in saturation, and the growth rate increases from slow to fast.
(3) In the process of sample preparation, the increase in soaking time will lead to the increase in dispersion of shale tensile strength, and may lead to the failure mode of the sample deviating from the Brazilian splitting theory model.
(4) In order to reduce the strength deterioration caused by hydration when studying the effect of moisture content on shale strength, the soaking time should be reduced as much as possible when preparing the samples.

Author Contributions: Conceptualization, L.X.; formal analysis, L.S.; funding acquisition, L.X.; investigation, L.X., B.H., Y.Z. and L.S.; methodology, L.X., B.H., Y.Z. and L.S.; project administration, L.X.; resources, L.X., B.H. and Y.Z.; supervision, L.X.; writing—original draft preparation, L.S.; writing—review and editing, L.X., B.H., Y.Z. and L.S.; data curation, B.H. and L.S.; validation, L.S.; visualization, L.S. All authors have read and agreed to the published version of the manuscript.

Funding: This research was funded by the "National Natural Science Foundation (No. U23B6004)" and the "Creative Project of Engineering Research Center of Alternative Energy Materials & Devices, Ministry of Education, Sichuan University (No. AEMD 202206)".

Data Availability Statement: The data that support the findings of this study are available from the corresponding author.

Conflicts of Interest: The authors declare no conflicts of interest.

References

1. Shen, D. China's Energy Problem and Alternative Solutions. *J. Contemp. China* **2001**, *10*, 717–722. [CrossRef]
2. Zou, C.; Pan, S.; Hao, Q. On the connotation, challenge and significance of China's "energy independence" strategy. *Pet. Explor. Dev.* **2020**, *47*, 449–462. [CrossRef]
3. Li, Y.; Zhou, D.H.; Wang, W.H.; Jiang, T.X.; Xue, Z.J. Development of unconventional gas and technologies adopted in China. *Energy Geosci.* **2020**, *1*, 55–68. [CrossRef]
4. Geologists, P.O.A.A.; Division, M.E. Unconventional Energy Resources: 2015 Review. *Nat. Resour. Res.* **2015**, *24*, 443–508.
5. Sieminski, A.; EIA Administrator. *International Energy Outlook*; U.S. Energy Information Administration: Washington, DC, USA, 2016.
6. Wang, J.S.; Ryan, D.; Anthony, E.J. Reducing the greenhouse gas footprint of shale gas. *Energy Policy* **2011**, *39*, 8196–8199. [CrossRef]
7. Mayerhofer, M.J.; Lolon, E.P.; Warpinski, N.R.; Cipolla, C.L.; Walser, D.; Rightmire, C.M. What Is Stimulated Reservoir Volume? *SPE Prod. Oper.* **2010**, *25*, 89–98. [CrossRef]
8. Wang, D.B.; Taleghani, A.D.; Yu, B. Height Effect on Interactions between the Hydraulic Fracture and Natural Fractures. *Geofluids* **2022**, *2022*, 4642326. [CrossRef]
9. Hubbert, M.K.; Willis, D.G. Mechanics of Hydraulic Fracturing. *Trans. Am. Inst. Min. Metall. Eng.* **1957**, *210*, 153–168. [CrossRef]
10. Haimson, B.; Fairhurst, C. Initiation and Extension of Hydraulic Fractures in Rocks. *Soc. Pet. Eng. J.* **1967**, *7*, 310–318. [CrossRef]
11. Ito, T.; Hayashi, K. Physical background to the breakdown pressure in hydraulic fracturing tectonic stress measurements. *Int. J. Rock Mech. Min. Sci. Geomech. Abstr.* **1991**, *28*, 285–293. [CrossRef]

12. Zhao, W.Z.; Hu, S.Y.; Hou, L.H. Connotation and strategic role of in-situ conversion processing of shale oil underground in the onshore China. *Pet. Explor. Dev.* **2018**, *45*, 563–572. [CrossRef]
13. Li, H.; Lai, B.; Liu, H.H.; Zhang, J.L.; Georgi, D. Experimental Investigation on Brazilian Tensile Strength of Organic-Rich Gas Shale. *SPE J.* **2016**, *22*, 148–161. [CrossRef]
14. Zheng, D.; Miska, S.; Ziaja, M.; Zhang, J. Study of anisotropic strength properties of shale. *AGH Drill. Oil Gas* **2019**, *36*, 93–112. [CrossRef]
15. Zheng, D.; Ozbayoglu, E.; Miska, S.; Zhang, J. Experimental Study of Anisotropic Strength Properties of Shale. In Proceedings of the 57th US Rock Mechanics/Geomechanics Symposium, Atlanta, GA, USA, 25–28 June 2023.
16. Teng, J.Y.; Tang, J.X.; Zhang, C. Experimental study on tensile strength of layered water-bearing shale. *Rock Soil Mech.* **2018**, *39*, 1317–1326.
17. Roy, D.G.; Singh, T.N.; Kodikara, J.; Ratan, D. Effect of Water Saturation on the Fracture and Mechanical Properties of Sedimentary Rocks. *Rock Mech. Rock Eng.* **2017**, *50*, 2585–2600.
18. Babets, D.V.; Kovrov, O.S.; Moldabayev, S.K.; Tereschuk, R.M. Impact of water saturation effect on sedimentary rocks strength properties. *Nauk. Visnyk Natsionalnoho Hirnychoho Universytetu* **2020**, *2020*, 76–81. [CrossRef]
19. Hasbollah, A.Z.D.; Mohamad, T.E.; Hezmi, A.M.; Abdullah, R.A.; Yunus, N.Z.M.; Mustaffar, M.; Jusoh, S.N.; Azmi, M.; Ramli, M.H.; Taib, A.M. Comparison study on the strength index of tropical shale and sandstone influenced by moisture content. In Proceedings of the IOP Conference Series: Materials Science and Engineering, Kuala Lumpur, Malaysia, 27–28 February 2019.
20. Lin, Y.F.; Guan, F.J.; Hu, H.Y. Adsorption Characteristics of Water-bearing Shale in Longmaxi Formation, Eastern Sichuang. *J. Xi'an Shiyou Univ. (Nat. Sci. Ed.)* **2019**, *34*, 21–25.
21. Li, W.T.; Pour-Ghaz, M.; Castro, J.; Weiss, J. Water absorption and critical degree of saturation relating to freeze-thaw damage in concrete pavement joints. *J. Mater. Civ. Eng.* **2012**, *24*, 299–307. [CrossRef]
22. Hatheway, W.A. The Complete ISRM Suggested Methods for Rock Characterization, Testing and Monitoring: 1974–2006. *Environ. Eng. Geosci.* **2009**, *15*, 47–48. [CrossRef]
23. Xue, H.Q.; Zhou, S.W.; Jiang, Y.L.; Zhang, F.D.; Dong, Z.; Guo, W. Effects of hydration on the microstructure and physical properties of shale. *Pet. Explor. Dev. Online* **2018**, *45*, 1146–1153. [CrossRef]
24. Wang, Q.; Lyu, C.; Cole, R.D. Effects of hydration on fractures and shale permeability under different confining pressures: An experimental study. *J. Pet. Sci. Eng.* **2019**, *176*, 745–753. [CrossRef]
25. Liu, J.; Bai, X.; Derek, E. Evolution of pore systems in low-maturity oil shales during thermal upgrading—Quantified by dynamic SEM and machine learning. *Pet. Sci.* 2024, *in press*. [CrossRef]
26. Liu, J.; Yao, Y.B.; Elsworth, D. Morphological complexity and azimuthal disorder of evolving pore space in low-maturity oil shale during in-situ thermal upgrading and impacts on permeability. *Pet. Sci.* 2024, *in press*. [CrossRef]
27. Huang, Z.; Zhang, Y.; Xie, L.Z.; Zhao, P.; He, B.; Ren, L. Comparative study of porosity test methods for shale. *Arab. J. Geosci.* **2020**, *13*, 94. [CrossRef]
28. Li, D.; Wong, L.N.Y. The Brazilian Disc Test for Rock Mechanics Applications: Review and New Insights. *Rock Mech. Rock Eng.* **2012**, *46*, 269–287. [CrossRef]
29. Dodge, Y. *The Concise Encyclopedia of Statistics*; Springer Science & Business Media: New York, NY, USA, 2008; pp. 95–267.
30. Yang, Z.P.; He, B.; Xie, L.Z.; Li, C.B.; Wang, J. Strength and failure modes of shale based on Brazilian test. *Rock Soil Mech.* **2015**, *36*, 3447–3455.
31. Van, E.E.M. The mechanisms of strength reduction due to moisture in coal mine shales. *Int. J. Rock Mech. Min. Sci. Geomech. Abstr.* **1976**, *13*, 61–67.
32. Chalmers, G.R.L.; Bustin, R.M.; Power, I.M. Characterization of gas shale pore systems by porosimetry, pycnometry, surface area, and field emission scanning electron microscopy/transmission electron microscopy image analyses: Examples from the Barnett, Woodford, Haynesville, Marcellus, and Doig units. *AAPG Bull.* **2012**, *96*, 1099–1119.
33. Zhou, T.; Zhang, S.C.; Yang, L.; Ma, X.F.; Zou, Y.S.; Lin, H. Experimental investigation on fracture surface strength softening induced by fracturing fluid imbibition and its impacts on flow conductivity in shale reservoirs. *J. Nat. Gas Sci. Eng.* **2016**, *36*, 893–905. [CrossRef]
34. Liu, X.J.; Zeng, W.; Liang, L.X.; Xiong, J. Experimental study on hydration damage mechanism of shale from the Longmaxi formation in southern Sichuan basin, China. *Petroleum* **2016**, *2*, 54–60. [CrossRef]
35. Liang, L.; Xiong, J.; Liu, X. Experimental study on crack propagation in shale formations considering hydration and wettability. *J. Nat. Gas Sci. Eng.* **2015**, *23*, 492–499. [CrossRef]
36. Yu, H.T.; Ding, Y.; Liu, Y.M.; Peng, M.; Liang, L.X.; Yu, X.L. A dynamical spontaneous imbibition model for shale considering hydration damage. *Pet. Drill. Tech.* **2023**, *51*, 139–148.
37. Qian, B.; Zhu, J.H.; Yang, H.; Liang, X.; Yin, C.B.; Shi, X.Z.; Li, D.Q.; Li, J.L.; Fang, H. Experiments on shale reservoirs plugs hydration. *Pet. Explor. Dev.* **2017**, *44*, 615–621. [CrossRef]
38. She, H.C.; Liu, S.Q.; Zhang, D.C.; Luo, K.; Hu, Z.Q. Research on Shale Hydration Disturbance Damage Test and Statistical Damage Model. *Chin. J. Undergr. Space Eng.* **2023**, *19*, 602–609.
39. Li, J.P.; Lian, M.J. *Mining Rock Mechanics*, 2nd ed.; Metallurgical Industry Press: Beijing, China, 2011; p. 57.

40. Shadizadeh, R.S.; Moslemizadeh, A.; Dezaki, S.A. A novel nonionic surfactant for inhibiting shale hydration. *Appl. Clay Sci.* **2015**, *118*, 74–86. [CrossRef]
41. Chao, Z.M.; Wang, H.L.; Xu, W.Y.; Jia, C.J.; Zhao, K. A rapid method for preparing rock samples with different water saturation levels. *Rock Soil Mech.* **2018**, *39*, 1109–1114.

energies

MDPI

Article

Insights into the Co-Exploration Potential of Gas in the Shale and Tight Sandstone of the Lower Silurian Formation in the Gongtan Syncline Area in Southeastern Sichuan Basin, SW China

Shengxiu Wang [1,2,3], Ye Zhang [1,2], Wei Wang [1,2,3,*], Yang Yang [4], Qiaoli Wang [1,2,3], Chuan Yu [1,2,3], Difei Zhao [5], Chunlin Zeng [1,2,3] and Yao Xu [1,2,3]

1 National and Local Joint Engineering Research Center of Shale Gas Exploration and Development, Chongqing Institute of Geology and Mineral Resources, Chongqing 401120, China; xiuxiu341@163.com (S.W.); zhangye_sg@vip.163.com (Y.Z.); wql19821031@163.com (Q.W.); yuchuan03041512@aliyun.com (C.Y.); ceng8623@189.cn (C.Z.); xuyao123s@163.com (Y.X.)
2 Key Laboratory of Shale Gas Exploration, Ministry of Natural Resources, Chongqing Institute of Geology and Mineral Resources, Chongqing 401120, China
3 Chongqing Huadi Resources and Environment Technology Co., Ltd., Chongqing 401120, China
4 China National Petroleum Corporation Changqing Oilfield Branch Fifth Gas Production Plant, Xi'an 710016, China; yangyang2_cq@petrochina.com.cn
5 Artificial Intelligence Research Institute, China University of Mining and Technology, Xuzhou 221000, China; difei.zhao@cumt.edu.cn
* Correspondence: cqdyysg_ww@126.com

Abstract: This work aims to explore the Lower Silurian shale gas and tight sandstone gas accumulation conditions in the Gongtan Syncline, southeastern Sichuan Basin. The sedimentary environment, organic geochemical characteristics, reservoir characteristics, gas content, and preservation conditions of the reservoir were comprehensively analyzed. The results show that the Wufeng–Longmaxi Formation shale formed in a deep-water shelf characterized by a large thickness (50–70 m), appropriate total organic carbon content (0.5–5.47%), high maturity (2.38%), high brittle mineral content (67.10%), and large gas content (0.71–1.64 m^3/t), and the formations show the good resource potential of the shale gas. The Xintan Formation formed in a lower shore phase, and the tight sandstone is locally developed with a small thickness. The Xiaoheba Formation formed in an upper-middle shore phase, and the tight sandstone is stably distributed with large thicknesses. The porosity and permeability of the two sets of sandstone are small and some natural fractures are developed in the sandstone, but the fracture filling degree is higher. The results of well logging show that there are abnormally high values of total hydrocarbon in both the Xintan Formation and Xiaoheba Formation; this indicates that tight sandstone gas is developed in the Lower Silurian strata. A comprehensive study indicates that the Lower Silurian of the Gongtan Syncline has the geological conditions for the formation of shale gas and tight sandstone gas, which are the "Two gases" with good co-exploration prospects.

Keywords: Gongtan Syncline; Lower Silurian; shale gas; tight sandstone gas; gas co-exploration

1. Introduction

With the continuous growth in global energy demand and the transition toward a low-carbon society, unconventional natural hydrocarbons have garnered increasing attention from nations worldwide [1–5]. Shale gas and tight sandstone gas have become focal points of research and exploration for unconventional natural gas resources in China. After years of exploration and development, these resources have become a tangible and essential component of China's energy portfolio [6–8]. In recent years, the Chongqing region, as a part of the southeastern Sichuan Basin, has emerged as a primary area for shale gas exploration in China. Breakthroughs have been achieved in the Wufeng–Longmaxi Formation in areas such as Fuling, Yuxi, Pengshui, and Nanchuan, confirming the substantial resource potential of this shale formation in the Chongqing region [9–12]. Meanwhile, previous studies have also indicated that the Sichuan Basin and surrounding areas harbor abundant tight sandstone gas resources, primarily within the Triassic (Xujiahe Formation), Jurassic, and Ordovician (Hanjiadian Formation and Xiaoheba Formation) strata [13–15]. Further, in the Chongqing and adjacent regions, drilling activities targeting the Ordovician Hanjiadian Formation and Xiaoheba Formation have demonstrated active oil and gas shows, with multiple wells achieving industrial gas flows [16,17], underscoring the promising exploration potential for tight sandstone gas in the Ordovician strata of the Chongqing region. However, a significant suite of tight sandstone reservoirs has been identified within the Ordovician strata in the southeastern region of the Sichuan Basin [18,19], but there has been limited foundational research into whether the southeastern region of the Sichuan Basin possesses the geological conditions conducive to the development of tight sandstone gas in the Ordovician strata.

In response to the coexistence of multiple oil and gas resources in the same region, scholars have proposed various methods for joint exploration and development of multiple resources [20–23]. These approaches not only help in cost-saving during exploration, but also ensure the comprehensive utilization of resources. As of now, successful examples of unconventional gas joint development have been seen around the world. For instance, in Colorado's Piceance Basin, deep coalbed methane and tight sandstone gas have been jointly extracted in a pioneering experiment [24,25]. In Wyoming's Wind River Basin, pressure-parted joint production of coalbed methane and sandstone gas has yielded individual well daily production rates of tens of thousands of cubic meters [26]. Similarly, in Canada's Horseshoe Canyon, coalbed methane along with the overlying and underlying sandstone and shale have been exploited jointly, resulting in higher gas production compared to single-layer extraction methods [27]. In China, regions with coexisting coalbed methane, shale gas, and tight sandstone gas are abundant, and in recent years, several studies have been conducted focusing on the characteristics of unconventional natural gas coexistence, resource potential assessment, and exploration and development strategies. For example, based on the research on unconventional natural gas accumulation conditions in the Qinshui Basin coal series, Li et al. established a model for the coexistence of coalbed methane, shale gas, and tight sandstone gas, indicating promising prospects for the joint exploration and development of these "Three gases" in the region [28]. Cao Daiyong et al. studied the geological conditions of unconventional natural gas accumulation in the coal series on the western margin of the Ordos Basin and established a model for the occurrence of coal-based unconventional natural gas [29]. Zhong Jianhua et al., based on their research on the gas-bearing seal characteristics of the coal series on the eastern margin of the Ordos Basin, proposed a model for the coexistence of tight sandstone gas, shale gas, and coalbed gas [30]. Zhou Yitong et al. conducted a coal series of "Three gases" joint production engineering practices in the Songhe Oilfield in western Guizhou, revealing promising development prospects for the "Three gases" in that area [31]. Previous comprehensive exploration and evaluation studies on the unconventional natural gas in China have often focused on coal-based unconventional gas, with limited research on the coexistence and feasibility of marine unconventional gas. However, in recent years, discoveries of coexisting shale gas and tight sandstone gas have been made in the marine strata of southern China [32,33], suggesting that southern marine strata also possess multiple types of unconventional gas accumulation conditions and configuration relationships, potentially enabling the joint exploration and exploitation of multiple gases. Nevertheless, research evaluating the joint exploration potential of shale gas and tight sandstone gas in the marine strata of southern China has yet to be undertaken.

In recent years, extensive exploration and evaluation of the shale gas in the Wufeng–Longmaxi Formation has been carried out in the southeastern Sichuan Basin. In 2015, during drilling operations at the GD 1 well located in the Gongtan Syncline, besides encountering favorable gas shows in the target Wufeng–Longmaxi Formation layers, gas anomalies were also detected in the overlying Xintan Formation tight sandstone. This discovery represents the first identification of tight sandstone gas in the Silurian strata in the southeastern marginal area of the Sichuan Basin [33], suggesting the potential coexistence of shale gas and tight sandstone gas in the Silurian strata of this region. Therefore, this study aims to analyze the fundamental conditions for the formation of Silurian shale gas and tight sandstone gas in the southeastern region of the Sichuan Basin, elucidate the unconventional natural gas resource potential in the Gongtan Syncline, and provide data support and guiding recommendations for the joint exploration and development of shale gas and tight sandstone gas in the study area.

2. Geological Settings and Sample Collection

2.1. Geological Background

The Gongtan Syncline is situated at the border between the southeastern region of Chongqing and the northern region of Guizhou (Figure 1), and it is a geological unit within the Yangtze Paraplatform (Level I), consisting of the Shangyitai Intra-Basin Depression (Level II) and the Yudongnan Depression and Fold Belt (Level III). The exposed parts of the syncline in the southeastern Sichuan Basin are divided into the northern and central segments, with an exposed length of approximately 20 km. The syncline comprises sedimentary strata from the Neoproterozoic era to the present, and it is dominated by Paleozoic formations. The axis of the syncline is primarily composed of the Lower Triassic Jialingjiang Formation, with Triassic, Permian, Devonian, and Ordovician formations exposed on both flanks. The Gongtan Syncline was mainly formed during the Late Yan Mountain and Himalayan periods. Despite undergoing extensive geological evolution, it exhibits good structural continuity and simplicity, trending in a north–northeast direction. The limbs of the syncline have steep dips, ranging from 30–86° on the northwest limb and 10–65° on the southeast limb. Faults are relatively rare and mostly occur near the fold axis, with short extensions (Figure 2).

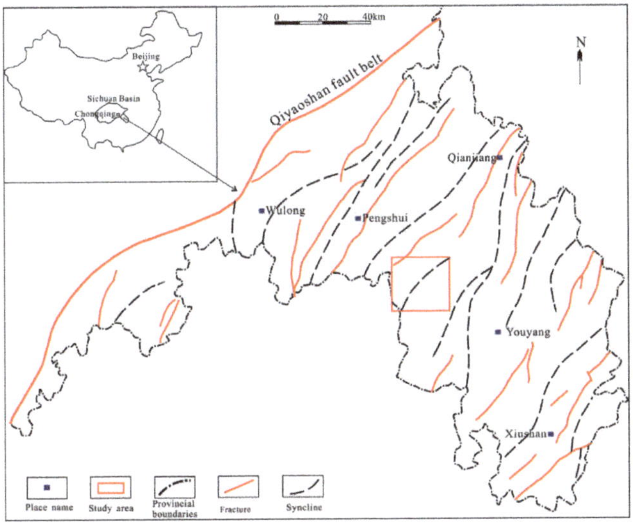

Figure 1. Regional location of the study area.

The Silurian strata in the southeastern Sichuan Basin, where the study area is located, are widely distributed, particularly in the lower and middle series. The upper part of the middle series and the upper series are absent within the regional context. From bottom to top, the stratigraphy consists of the Longmaxi Formation, Xintan Formation, Xiaoheba Formation, and Hanjiadian Formation. The Longmaxi Formation and the underlying

Wufeng Formation are characterized by dark shale. Reservoirs of Silurian clastic rocks are mainly distributed in the Middle Silurian Hanjiadian Formation, the Lower Silurian Xiaoheba Formation, and the Xintan Formation. However, the Hanjiadian Formation suffers from severe erosion and extensive exposure within the Gongtan Syncline.

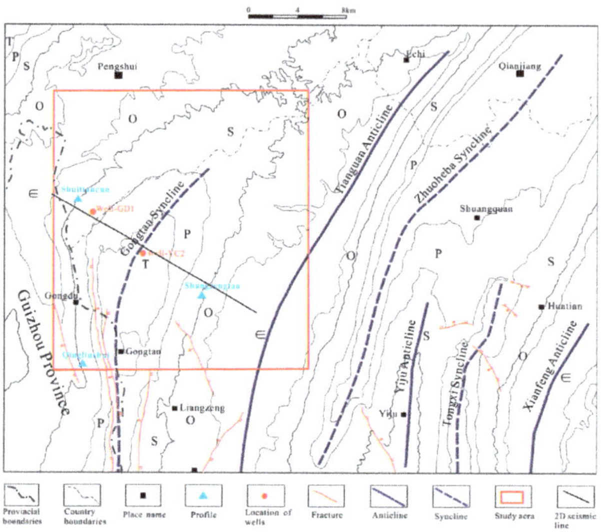

Figure 2. Regional structure diagram of the Gongtan Syncline. The arrow refers to the tendency of faults.

2.2. Sampling and Methodology

Shale samples from the Wufeng and Longmaxi Formations were collected from the cores of shale gas Wells GD 1 and YC 2, as well as from three outcrop profiles within the study area, totaling 100 samples. Additionally, tight sandstone samples from the Xintan and Xiaoheba Formations were collected from Well GD 1, totaling 15 samples. The collected shale samples were subjected to a series of analyses, including TOC, R_o, maceral composition, thin section analysis, whole-rock X-ray diffraction quantitative analysis, scanning electron microscopy, porosity and permeability measurements, and gas desorption. Tight sandstone samples underwent thin section analysis, as well as porosity and permeability measurements.

In this study, the organic geochemical characteristics of shale were characterized using three parameters: TOC, R_o, and organic matter type. The German Leica Instrument Company's DM4500P optical microscope was employed to measure the R_o and maceral composition, assessing the thermal maturity and organic matter type of the shale. The corresponding analysis methods followed the standards of the Chinese petroleum and natural gas industry, specifically SY/T 5124-2012 [34] and SY/T 5125-2014 [35]. To determine the total organic carbon content of the samples, a China HaiCheng Rock-Eval VIII instrument (produced by HaiCheng Petrochemical Instrument Factory) equipped with a flame ionization detector was employed. The corresponding testing procedures adhered to the Chinese national standard GB/T 18602-2012 [36].

The reservoir characterizations of the shale and tight sandstone in this study primarily involved analyzing mineral composition characteristics, pore structure features, and porosity–permeability characteristics. Mineral composition characteristics were analyzed using X-ray diffraction (XRD) technology with a Bruker D8 Advance X-ray diffractometer from Bruker Corporation, Berlin, Germany, following the guidelines of the Chinese petroleum and natural gas industry standard SY/T 5163-2010 [37]. The pore structure was determined using a JSM-6610LV scanning electron microscope (SEM) from JEOL, Beijing, Chian and an X-max energy dispersive spectrometer from Oxford Instruments, Shanghai, Chian. Corresponding analysis methods adhered to the Chinese national standards GB/T

18295-2001 [38] and GB/T 17359-2012 [39]. Porosity and permeability were determined using an AP-609 porosity–permeability testing instrument from Coretest Systems following the procedures outlined in the Chinese national standard GB/T 29172-2012 [40].

An integrated analysis of sedimentary environments was conducted by combining the field outcrop data, drilling logs, well-logging data, and thin section analysis results. The thin section analysis was primarily aimed at identifying rock types and providing data support for the depositional facies division of the target layers. The experiments utilized a Leica Scope.A1 polarizing microscope from Leica Instrument Company, Beijing, China, following the guidelines of national standard GB/T 17412.1~3-1998 [41]. Shale gas content testing was performed using an on-site gas content testing instrument developed by the Chongqing Institute of Geology and Mineral Resources. The test results were primarily used to evaluate the shale gas resource potential, with on-site testing methods following the Chinese national standard GB/T 19559-2008 [42].

3. Gas Accumulation in Shale and Tight Sandstone

3.1. Favorable Geological Condition for Shale Gas

3.1.1. Sedimentary Environment

Previous studies have indicated that the Silurian strata in the southeastern part of the Sichuan Basin were deposited in a shallow marine shelf environment [43,44]. Based on a comprehensive analysis of drilling, outcrop data, and thin-section identification, along with previous research findings, the Wufeng–Longmaxi Formation in the southeastern of the Sichuan Basin can be vertically subdivided into two subfacies and four microfacies sedimentary types. The two subfacies are a shallow-water continental shelf and a deep-water continental shelf. The four microfacies are deep-water argillaceous continental shelf facies, shallow-water mixed continental shelf facies, shallow-water argillaceous continental shelf facies, and shallow-water sandy continental shelf facies. Organic-rich shale mainly develops in the deep-water mudstone shelf sub-facies. On a planar scale, most areas of the southeastern Sichuan Basin are situated in a deep-water shelf environment, with only parts of Youyang County and Xiushan County in a shallow-water shelf environment. The Wufeng–Longmaxi Formation in the Gongtan Syncline is located within a deep-water shelf environment (Figure 3), which represents a favorable sedimentary environment for shale gas development.

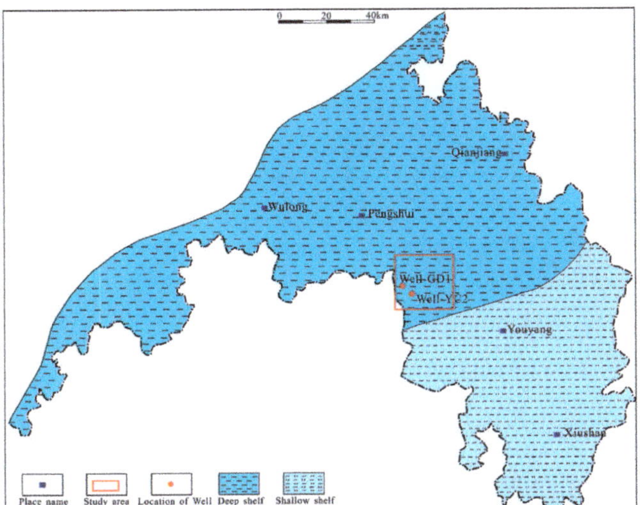

Figure 3. Schematic diagram of the Wufeng–Longmaxi Formation in the southeastern Sichuan Basin (modified by reference [45]).

3.1.2. Hydrocarbon Source Rock Conditions

The mud shale of the Wufeng–Longmaxi Formation represents a regionally excellent hydrocarbon source rock [45]. In partial areas of the southeastern Sichuan Basin, this stratigraphic unit has undergone erosion, resulting in significant variations in sediment thickness, ranging from 30 to 120 m, with a gradual thinning trend from northwest to southeast [46]. Drilling and outcrop data reveal that, within the Gongtan Syncline, the distribution of the Wufeng–Longmaxi Formation mud shale is relatively stable, with thickness primarily ranging between 50 to 70 m. Specifically, in the GD 1 well, the thickness ranges from 1180 to 1259 m (a total of 79 m) (refer to Figure 4a), while in the YC 2 well, the thickness ranges from 2104 to 2182 m (a total of 78 m) (refer to Figure 4b). An organic geochemical analysis of 100 samples from drilling and outcrop materials indicates that the organic matter in the Wufeng–Longmaxi Formation shale is predominantly Type I and Type II_1, with organic carbon content ranging mainly from 0.21% to 4.91% and averaging at 1.10% (refer to Figure 5a). The organic matter maturity is relatively high, with vitrinite reflectance (R_o) ranging from 1.19% to 2.39% and averaging at 2.18% (refer to Figure 5b), indicating an evolution stage of high to over maturity. The mud shale has reached the peak generation period, facilitating the full generation of shale gas. Consequently, the mud shale of the Wufeng–Longmaxi Formation in the southeastern Sichuan Basin serves as a favorable hydrocarbon source rock, providing conducive conditions for the generation of Lower Silurian shale gas and tight sandstone gas in the region.

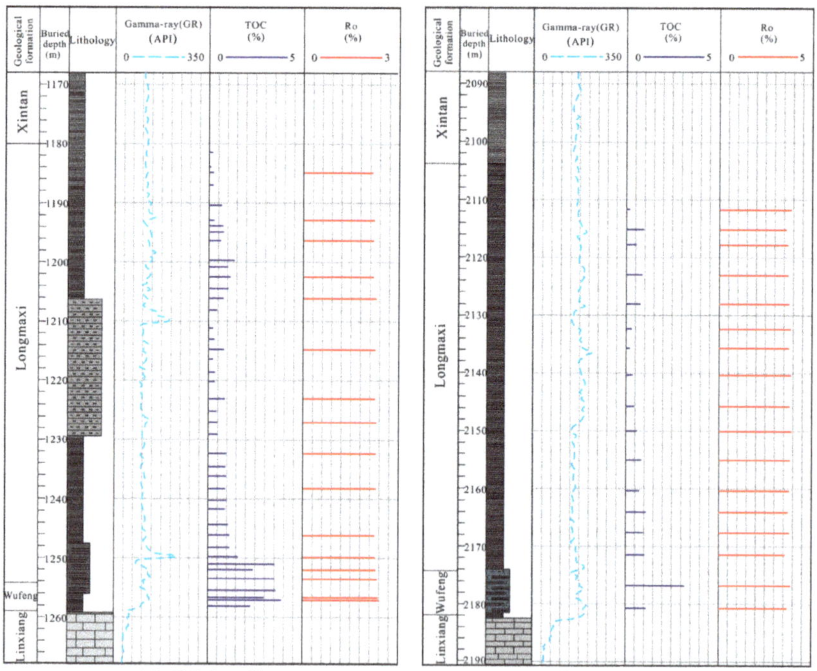

(a). GD 1 well (b). YC 2 well

Figure 4. Columnar diagram of the organic geochemical characteristics of the Wufeng–Longmaxi Formation shales in the study area.

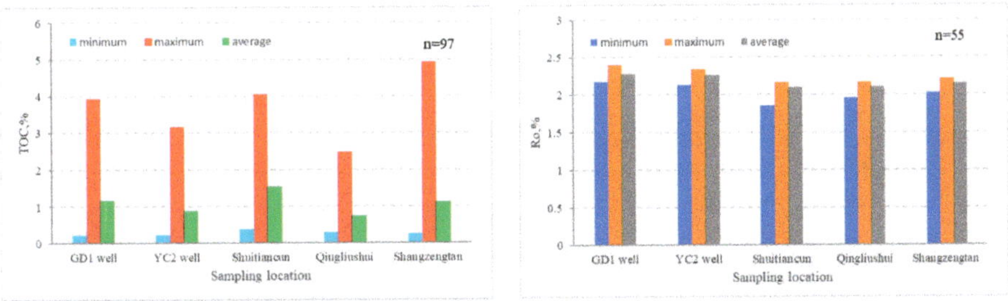

(**a**). TOC (**b**). Ro

Figure 5. Histogram of TOC (**a**) and R_o (**b**) for the Wufeng–Longmaxi Formation shales in the study area.

3.1.3. Reservoir Characteristics

The shale of the Wufeng–Longmaxi Formation constitutes a "source-reservoir unity," serving as both the hydrocarbon source rock and the reservoir for shale gas [47]. Results from an X-ray diffraction (XRD) analysis of 36 drilling and outcrop samples within the Gongtan Syncline indicate that the predominant mineral composition of the Wufeng–Longmaxi Formation shale comprises detrital minerals and clay minerals (refer to Figure 6), with minor quantities of carbonate and pyrite, among other minerals. The shale contains a high content of brittle minerals such as quartz, feldspar, and carbonate, averaging at 67.10%, which is conducive to reservoir fracturing and enhancement. The scanning electron microscope (SEM) results reveal that the matrix pore types in the shale of the Wufeng–Longmaxi Formation in the study area are primarily organic matter pores and interlayer clay mineral pores, with well-developed structural fractures that provide favorable storage space for shale gas enrichment. The porosity and permeability of 47 shale core samples from two wells were tested by the thermal pulse-decay method. The results showed that the porosity of the samples ranged from 0.19 to 5.41%, with an average of 1.21%, and the permeability ranged from 0.0001 to 0.2223 mD, with an average of 0.0082 mD. In general, the Wufeng–Longmaxi Formation in the study area belongs to ultra-low porosity and ultra-low permeability reservoirs.

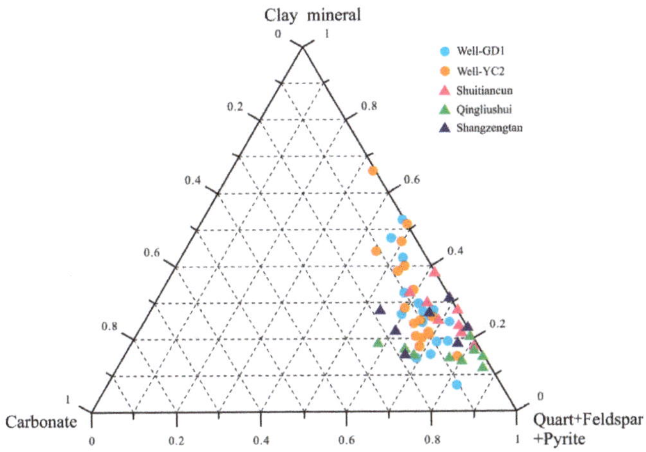

Figure 6. Mineral triangular diagram of the shale of the Wufeng–Longmaxi Formation in the Gongtan Syncline.

3.1.4. Gas Content

The gas content of shale is a key parameter for evaluating the potential of shale gas reservoirs, and it is crucial for shale gas evaluation, favorable area selection, resource estimation, and calculation [48]. Shale gas content parameters are typically obtained through on-site desorption or isothermal adsorption experiments [49]. In this study, the gas content parameters were primarily obtained using on-site desorption methods, where the desorbed gas quantity was measured using instruments at the drilling site. Statistical analysis of the on-site desorption experiment data from the GD 1 and YC 2 wells within the Gongtan Syncline showed that the Wufeng–Longmaxi Formation shale exhibits good gas content. Vertically, the gas content tends to increase with depth. Specifically, the on-site desorption gas content of the core samples from the GD 1 well ranged from 0.42 to 3.02 m^3/ton, averaging 1.64 m^3/ton (Figure 7a). For the YC 2 well, the on-site desorption gas content ranged from 0.11 to 2.39 m^3/ton, averaging 0.71 m^3/ton (Figure 7b). These values are comparable to the on-site analyzed gas content of 0.90 m^3/ton at the Jiaoye-1 well in the Reefs Dam area [50], indicating the favorable shale gas resource potential of the Wufeng–Longmaxi Formation within the Gongtan Syncline.

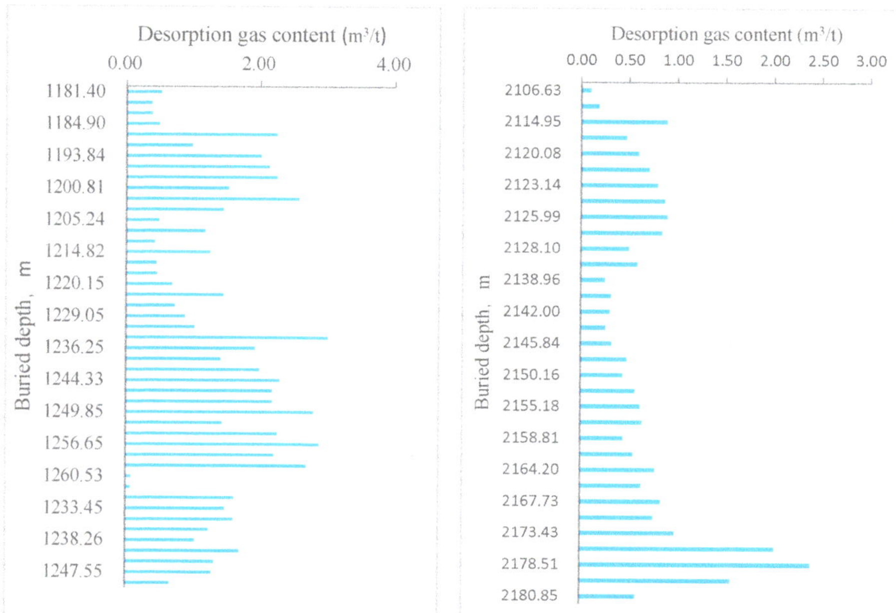

(**a**). GD 1 well. (**b**). YC 2 well.

Figure 7. Variation of the on-site desorption gas content with the buried depth of the Wufeng–Longmaxi Formation in the Gongtan Syncline.

3.2. *Favorable Geological Conditions for Tight Sandstone Gas*

3.2.1. Sedimentary Environment

Based on the analysis of drilling and logging data in the southeastern Sichuan Basin, the overlying Xintan Formation and Xiaoheba Formation above the Wufeng–Longmaxi Formation primarily exhibit three sedimentary subenvironments: marginal marine (including lower and middle-upper), offshore transition, and outer shelf (Figures 8 and 9). Sandstones are mainly deposited in the middle-upper marginal marine environment, with limited shoreface sand barriers in the lower marginal marine setting and more common occurrences of mudstone interbeds. In the offshore transition zone, there are occasional storm-induced sand barriers, but they are limited in distribution and scale. The Xintan Formation predominantly consists of the sand barriers and mud flats of the lower shoreface within the study area (Figure 8), with tight sandstones developed only in the sand barriers on the western flank of the syncline. The Xiaoheba Formation in the Gongtan Syncline mainly features coastal sand barriers in the middle-upper marginal marine setting

(Figure 9), with a relatively developed occurrence of tight sandstones. Therefore, the Xintan Formation and Xiaoheba Formation in the study area represent sedimentary environments conducive to the development of tight sandstones, providing the material basis for the formation of tight sandstone reservoirs in the region.

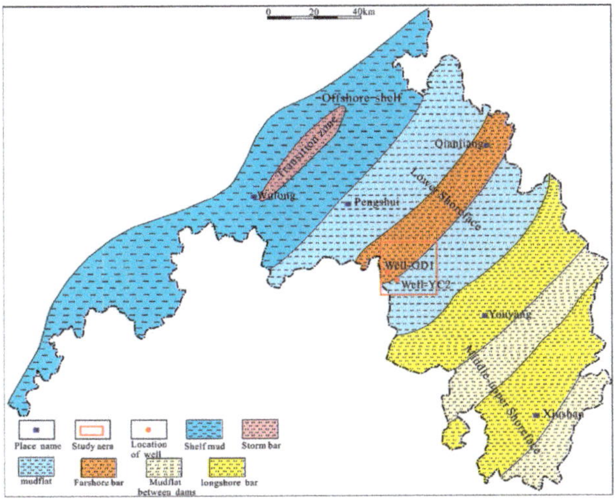

Figure 8. Schematic diagram of the sediments of the Xintan Formation in the southeastern Sichuan Basin (modified by reference [51]).

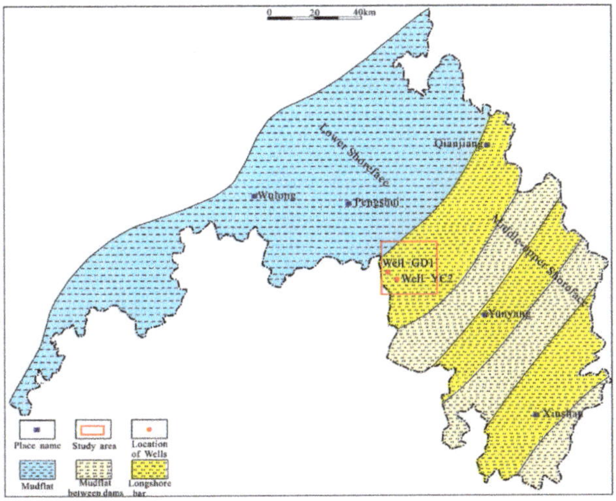

Figure 9. Schematic diagram of the sediments of the Xiaoheba Formation in the southeastern Sichuan Basin (modified by reference [52,53]).

3.2.2. Reservoir Characteristics

Analysis of the drilling and logging data from Wells GD 1 and YC 2 revealed the presence of discontinuous sand bodies in the middle-upper part of the Xintan Formation within the study area. These sand bodies exhibit unstable single-layer thicknesses ranging from 2 to 15 m, and they are primarily composed of calcareous silty sandstone. The lower part of the Xiaoheba Formation shows continuous and relatively thick-bedded sand bodies, with thicknesses ranging from 26 to 50 m. The lithology of these sand bodies mainly consists of fine-to-silty sandstone and calcareous silty sandstone, exhibiting coarser grain sizes compared to the mudstone–siltstone of the Xintan Formation. Physical property analyses of 15 core samples from the GD1 well (Figure 10) showed low porosity and permeability values for the middle-upper part of the Xintan Formation and the lower part of the Xiaoheba Formation. Porosity values ranged from 0.83% to 3.96%, with an average of 2.12%, while permeability ranged from 0.0015 to 0.0920 mD, with an average of 0.0227 mD. Interpretation of the logging data indicates that, even in the middle-upper Xintan and lower Xiaoheba Formations, the tight sandstone reservoirs exhibit extremely low porosity and low permeability characteristics (Table 1). In summary, the sandstone reservoirs within the Xintan and Xiaoheba Formations exhibit poor physical properties, and they are characterized as having extremely low porosity and low permeability tight sandstone reservoirs.

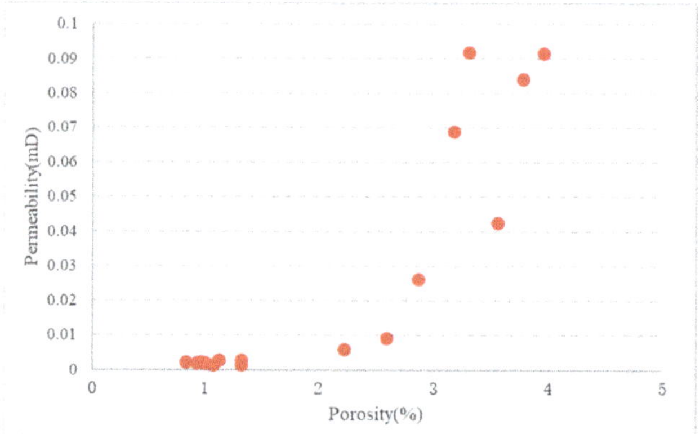

Figure 10. Porosity and permeability of the Lower Silurian tight sandstone in the GD 1 well.

Table 1. Reservoir petrophysical parameters of the tight sandstones from the Xintan and Xiaoheba formations. The value of porosity and permeability refers to the minimum-maximum and the average.

Well	Top Depth (m)	Basal Depth (m)	Porosity (%)	Permeability (mD)
GD 1	978	1030	0–6.448 2.731	0–0.414 0.077
YC 2	1670	1900	1.026–6.074 2.966	0–0.255 0.023

3.2.3. Gas Content

The existing drilling and logging data indicate that the two wells not only exhibit good gas indications within the syncline, but also show gas indications in the overlying Xintan and Xiaoheba formations. The gas logging curves from the GD 1 well revealed consistent gas shows from the lower-middle part of the Xintan Formation (around 1000 m) down to the base of the Wufeng Formation (Figure 11). In the middle section of the Xintan Formation (approximately 1010 m to 1055 m), gas anomalies were detected in silty mudstone and mudstone–siltstone intervals, with total hydrocarbon contents ranging from 1.651% to 12.183%. Logging interpretation of this section indicates the presence of gas-bearing zones, suggesting good gas saturation in the mudstone–siltstone of the Xintan Formation. At the YC 2 well, gas anomalies were identified in the lower part of the calcareous silty sandstone within the Xiaoheba Formation (1788–1796 m), with total hydrocarbon contents ranging from 0.196% to 7.384% (Figure 12). The logging interpretation of this section also confirmed the presence of gas-bearing zones in this interval. These findings from both wells encountering tight sandstone gas within the Xintan and Xiaoheba Formations further confirm the geological characteristic of coexisting "dual-gas" in the lower Silurian strata of the study area.

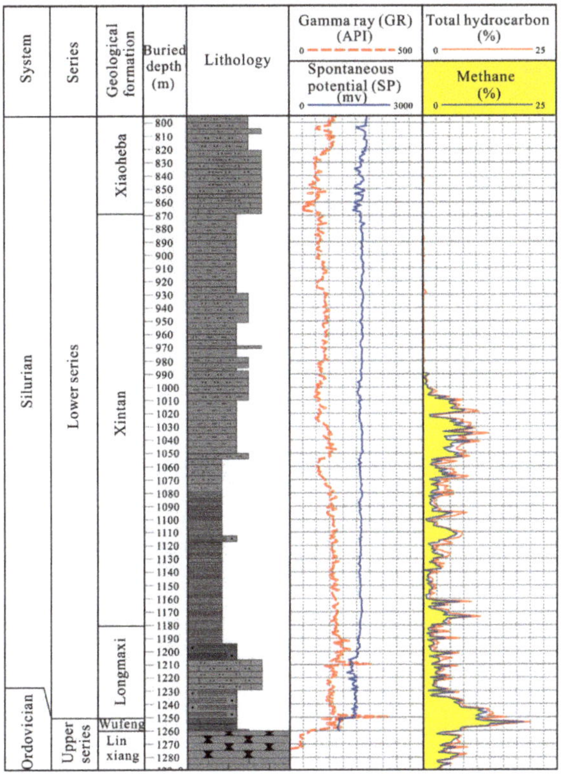

Figure 11. Histogram of the gas logging of the GD 1 well.

3.2.4. Hydrocarbon Migration Pathways

(1) Reservoir fractures and hydrocarbon accumulation

Based on the research results above, it is evident that the high mud content in the tight sandstones of the Xintan and Xiaoheba formations within the study area leads to relatively poor reservoir properties, which are unfavorable for the migration and accumulation of natural gas. Scholars have discovered that natural fractures developed within tight reservoirs can serve as pathways for oil and gas migration and as trapping locations, thereby improving the conditions for oil and gas accumulation in tight reservoirs [54,55]. Through core observations, fractures were identified within the gas-rich intervals of the mudstone–siltstone in the middle-upper part of the Xintan Formation in the GD 1 well (Figure 13). These fractures predominantly consist of low-angle and inclined fractures. The fracture density is relatively low, with an average longitudinal fracture density of only 0.11 fractures per meter, primarily consisting of small fractures. The statistical analysis of fracture filling indicates that most fractures are filled, with fully filled fractures accounting for 87.2% of all fractures. The imaging logging data from Well YC 2 (Figure 14) show that fractures within the tight sandstone of the Xiaoheba Formation mainly consist of high-resistance fractures and drilling-induced fractures, with minimal development of high-conductivity fractures. As a result, natural fractures are developed to a certain extent within the Lower Silurian tight sandstone formations in the study area, but these fractures are predominantly filled, which may partially affect the migration and accumulation of natural gas within the tight sandstone reservoirs.

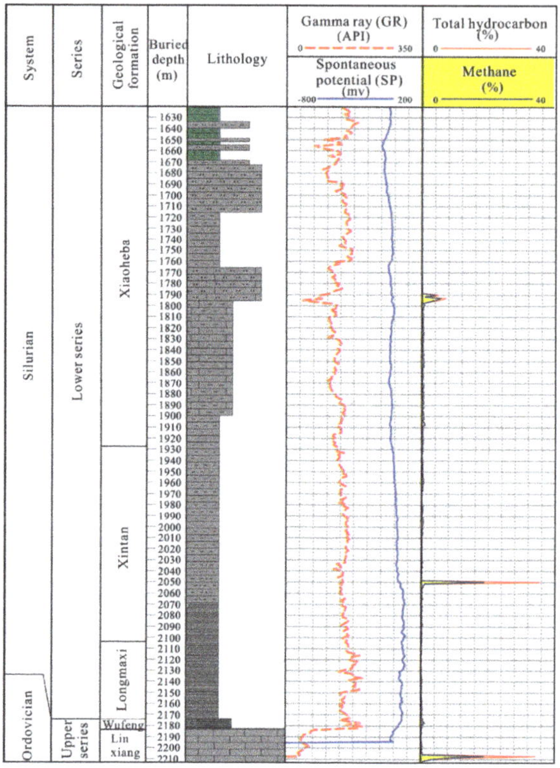

Figure 12. Histogram of the gas logging of the YC 2 well.

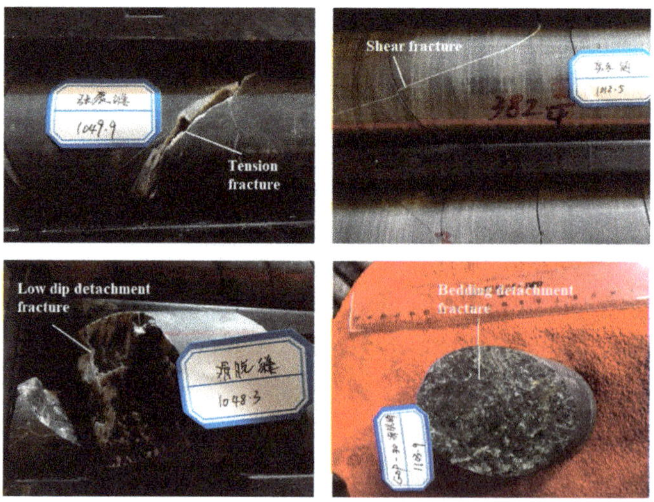

Figure 13. Fracture characteristics of the tight sandstone of the Xintan Formation in the GD 1 well.

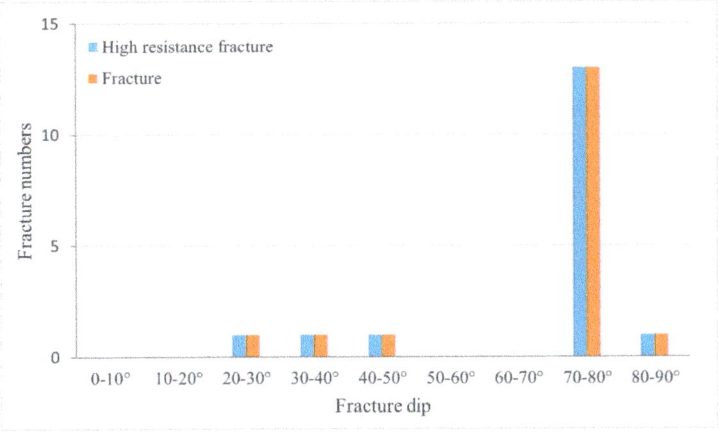

Figure 14. Dip angles of the natural fractures in the Xiaoheba Formation, Well YC 2.

(2) Fractures of the hydrocarbon source rock and hydrocarbon accumulation

For the tight sandstone gas reservoirs in the Xintan and Xiaoheba Formations within the study area, the organic-rich shale of the Upper Ordovician Wufeng Formation to the Lower Silurian Longmaxi Formation serves as the main source rock for hydrocarbons. Natural gas within shale layers require effective migration pathways for transport [56–58], whereas these shale layers exhibited characteristics of self-generation, self-storage, tightness, low porosity, and ultra-low permeability. The presence of natural fractures undoubtedly can provide crucial conditions for the migration of gas within shale layers, facilitating gas expulsion. Through core observations, it was observed that natural fractures are developed within the Wufeng–Longmaxi formations in the wells of GD 1 and YC 2 (Figure 15). These fractures predominantly consist of high-angle cutting and vertical fractures. Such fractures effectively connect different mechanical layers of the rock, forming efficient pathways for natural gas migration. Consequently, natural gas generated within the Wufeng–Longmaxi shale formations can migrate into the overlying Xintan and Xiaoheba formations of tight sandstones through these pathways.

(**a**). Core samples from the lower part of the Wufeng–Longmaxi Formation in Well GD 1.

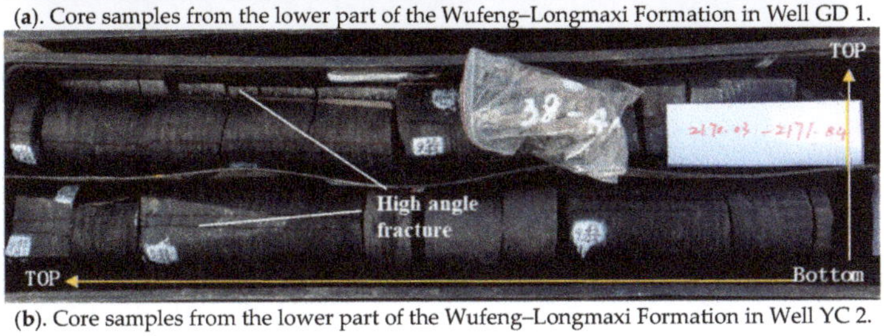

(**b**). Core samples from the lower part of the Wufeng–Longmaxi Formation in Well YC 2.

Figure 15. Fracture characteristics of the Wufeng–Longmaxi Formation in the Gongtan Syncline.

4. Gas Preservation Settings

The Wufeng–Longmaxi Formation shales are extensively distributed within the Gongtan Syncline at moderate burial depths. Overlying these formations are medium-thick layers of silty mudstone, muddy siltstone, and fine sandstone belonging to the Xintan and Xiaoheba formations, serving as both reservoirs of tight sandstone gas and caprocks for Wufeng–Longmaxi Formation shales. The Hanjiadian Formation overlies the Xintan and Xiaoheba Formations. Its lower portion predominantly consists of yellow and yellow-green shales interbedded with a purple-red shale. The middle section is composed of yellow-green, gray-green, and purple-red shales along with thick layers of siltstone. The upper part of the Hanjiadian Formation is characterized by yellow-gray and gray-green shales. The Hanjiadian Formation exhibits significant thickness, ranging from 703 to 783 m, and it is extensively distributed throughout the study area. It is characterized by low porosity, low permeability, and effective sealing properties. Consequently, the development of direct caprocks for shale gas and tight sandstone gas is well established, exhibiting favorable sealing characteristics.

Seismic interpretation results indicate that the Gongtan Syncline is a typical dish-shaped fold with a relatively weak structural deformation. Only a few faults with relatively small fault scales and spacings are present within the Xintan and Longmaxi formations, as well as underlying formations in the syncline. These faults do not breach the overlying layers of the Lower Silurian shale and tight sandstone reservoirs (Figure 16). Overall, the structural style of the Gongtan Syncline is straightforward, and the faults developed within the syncline have minimal impact on the preservation conditions of shale gas and tight sandstone gas, making it an advantageous area for unconventional oil and gas preservation in the southeastern part of the Sichuan Basin.

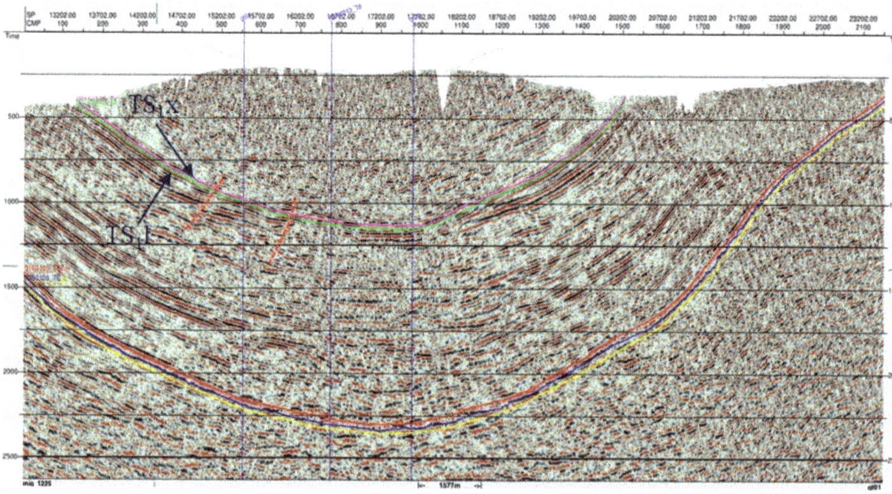

Figure 16. Structural profile of the Gongtan Syncline (Profile lines are shown in Figure 2). The red dotted lines represent faults in the Gongtan Syncline.

5. Conclusions

(1) Within the Gongtan Syncline, the Wufeng–Longmaxi Formation is situated in a deep-water continental shelf depositional environment, characterized by thick organic-rich shale deposits with high organic content, moderate thermal maturity, and good hydrocarbon generation potential. This provides a solid material basis for the accumulation of shale gas and tight sandstone gas. The Wufeng–Longmaxi Formation in the study area exhibits high desorption gas content from organic-rich shale, indicating good gas-bearing properties and significant potential for shale gas resources.

(2) The Xintan Formation primarily comprises offshore sandbars and nearshore mud flats of the lower shoreface, with an unstable distribution of tight sandstone layers. These tight sandstone layers have relatively small individual thicknesses, predominantly developing within the offshore sandbars on the western flank of the syncline. The Xiaoheba Formation is dominated by alongshore sandbars of the upper-middle shore phase, with well-developed tight sandstone that is characterized by thicker individual layer thickness and more stable distribution. Both sets of tight sandstone reservoirs exhibit low porosity, low permeability, and some natural fractures characterized by high fracture filling. These natural fractures may facilitate the extensive migration and accumulation of natural gas in the formations.

(3) The Lower Silurian formations within the Gongtan Syncline are at moderate burial depths, and they are characterized by overall underdeveloped internal faults and a relatively intact cap rock. The structural style is a broad and gentle syncline, providing favorable conditions for hydrocarbon preservation. The geological conditions for shale gas accumulation in the Wufeng–Longmaxi Formation within the syncline are favorable, offering conducive prospects for shale gas exploration. The accumulation conditions for tight sandstone gas in the Xiaoheba Formation are superior to those in the Xintan Formation. The tight sandstone gas in the Xintan Formation is locally accumulated due to the distribution of tight sandstones, which are primarily developing on the western flank of the syncline.

Author Contributions: Methodology, W.W., Y.Y., Q.W. and D.Z.; Validation, W.W.; Investigation, S.W. and C.Z.; Resources, Y.Z. and D.Z.; Writing—original draft, S.W.; Writing—review & editing, W.W.; Visualization, C.Y. and Y.X.; Project administration, Y.Z. All authors have read and agreed to the published version of the manuscript.

Funding: This study was financially supported by the Chongqing Natural Science Foundation (cstc2021jcyj-msxmX0624, CSTB2023NSCQ-LZX0036).

Data Availability Statement: The data used to support the findings of this study are available from the corresponding authors upon request.

Conflicts of Interest: Authors Shengxiu Wang, Wei Wang, Qiaoli Wang, Chuan Yu, Chunlin Zeng and Yao Xu were employed by the Chongqing Huadi Resources and Environment Technology Co., Ltd. Author Yang Yang was employed by the China National Petroleum Corporation Changqing Oilfield Branch Fifth Gas Production Plant. The authors declare that the research was conducted in the absence of any commercial or financial relationships that could be construed as potential conflicts of interest.

References

1. Jarvie, D.M.; Hill, R.J.; Ruble, T.E.; Pollastro, R.M. Unconventional shale-gas systems: The mississippian barnett shale of north-central Texas as one model for thermogenic shale-gas assessment. *AAPG Bull.* **2007**, *91*, 475–499. [CrossRef]
2. Marc Bustin, R. Shale gas and shale oil petrology and petrophysics. *Int. J. Coal Geol.* **2012**, *103*, 1–2. [CrossRef]
3. Liu, J.; Yao, Y.B.; Elsworth, D. Morphological complexity and azimuthal disorder of evolving pore space in low-maturity oil shale during in-situ thermal upgrading and impacts on permeability. *Pet. Sci.* 2024, *in press*. [CrossRef]
4. Wakamatsu, H.; Aruga, K. The impact of the shale gas revolution on the U.S. and Japanese natural gas markets. *Energy Policy* **2013**, *62*, 1002–1009. [CrossRef]
5. Zhou, G.N.; Yang, Z.; Zhu, R.K.; Zhang, G.S.; Hou, L.H.; Wu, S.T.; Shizhen, T.A.O.; Yuan, X.J.; Dong, D.Z.; Wang, Y.M.; et al. Progress in China's Unconventional Oil & Gas Exploration and Development and Theoretical Technologies. *Acta Geol. Sin.* **2015**, *89*, 979–1007.
6. Li, L.; Liu, J.; Wang, W.; Zhao, P.; Lu, Y. Variational Reservoir Characteristics of the Wufeng–Longmaxi Formation from Different Sedimentary Region and Its Implications in Southeastern Sichuan Basin, China. *ACS Omega* **2023**, *8*, 20684–20696. [CrossRef]
7. Zou, C.; Zhao, Q.; Cong, L.; Wang, H.; Shi, Z.; Wu, J.; Pan, S. Development progress, potential and prospect of shale gas in China. *Nat. Gas Ind.* **2021**, *41*, 1–14.
8. Jia, A.; Wei, Y.; Guo, Z.; Wang, G.; Meng, D.; Huang, S. Development status and prospect of tight sandstone gas in China. *Nat. Gas Ind.* **2022**, *42*, 83–92. [CrossRef]
9. Guo, T.L. Discovery and characteristics of the Fuling shale gas field and its enlightment and thinking. *Earth Sci. Front.* **2016**, *23*, 29–43.

10. He, X.P.; Zhang, P.X.; Fang, D.Z.; Mei, J.W.; He, G.S.; Lu, B. Production characteristics of normal pressure shale gas in Pengshui-Wulong area, southeast Chongqing. *Pet. Geol. Recovery Effic.* **2018**, *25*, 72–79.
11. Guo, X.S. Controlling factors on shale gas accumulations of Wufeng-Longmaxi Formations in Pingqiao shale gas field in Fuling area, Sichuan Basin. *Nat. Gas Geosci.* **2019**, *30*, 1–10. [CrossRef]
12. Zhang, S.R.; Dong, D.Z.; Liao, Q.S.; Sun, S.; Huang, S.; Guan, Q.; Zhang, C.; Guo, W.; Jiang, S.; Shi, P. Geological characteristics and resource prospect of deep marine shale gas in the southern Sichuan Basin. *Nat. Gas Ind.* **2021**, *41*, 35–45.
13. Ma, D.F. Evaluation of Silurian Tight Sandstone Reservoir in Jiannan Gas Field Area, West Hubei-East Chongqing. *Mar. Orig. Pet. Geol.* **2011**, *16*, 14–19.
14. Guo, T.L. Key controls on accumulation and high production of large non-marine gas fields in northern Sichuan Basin. *Pet. Explor. Dev.* **2013**, *40*, 139–149. [CrossRef]
15. Yang, G.; Li, G.; Li, N.; Chen, S.; Wang, H.; Xu, L. Hydrocarbon accumulation characteristics and enrichment laws of multi-layered reservoirs in the Sichuan Basin. *Nat. Gas Ind. B* **2017**, *4*, 81–90. [CrossRef]
16. Zhu, Z.J.; Zhang, Z.B. Depositional system evolution characteristic in the framework of sequences of Silurian and prediction of Favorable zones in eastern Sichuan-Western Hubei. *Sci. Technol. Eng.* **2016**, *16*, 13–21.
17. Zhang, W.K.; Shi, Z.J.; Tian, Y.M.; Wang, Y.; Hu, X.Q.; Li, W.J. Pore types and genesis of tight sandstone of Silurian Xiaoheba Formation in southeastern Sichuan Basin. *Lithol. Reserv.* **2021**, *33*, 10–19.
18. Ma, W.X.; Liu, S.G.; Huang, W.M.; Zhang, C.J.; Zeng, X.L. Reservoir rocks characters of Silurian and its unconventional gas prospection in Western Hubei—Eastern Chongqing. *J. Southwest Pet. Univ. Sci. Technol. Ed.* **2012**, *34*, 27–37.
19. Wang, L.J.; Zhang, W.J.; Hu, X.W.; Luo, Y.P.; Cheng, L.X. Pore structures of the Xiaoheba Formation tight sandstone reservoirs in southeastern Sichuan Basin, China. *J. Chengdu Univ. Technol.* **2019**, *46*, 70–79.
20. Liang, B.; Shi, Y.S.; Sun, W.J.; Liu, Q. Reservoir forming characteristics of "the three gases" in coal measure and the possibility of commingling in China. *J. China Coal Soc.* **2016**, *41*, 167–173.
21. Zhang, S.Q.; Liu, J.; Li, L.; Kassabi, N.; Hamdi, E. Petrophysical and Geochemical Investigation-Based Methodology for Analysis of the Multilithology of the Permian Longtan Formation in Southeastern Sichuan Basin, SW China. *Energies* **2024**, *17*, 766. [CrossRef]
22. Zou, C.N.; Yang, Z. Resource types, formation, distribution and prospects of coal-meas-ure gas. *Pet. Explor. Dev.* **2019**, *46*, 451–462. [CrossRef]
23. Zheng, L.H.; Wei, P.F.; Zhang, Z.; Nie, S.S.; Lou, X.Q.; Cui, K.X.; Fu, Y.W. Joint exploration and development: A self-salvation road to sustainable development of unconventional oil and gas resources. *Nat. Gas Ind.* **2017**, *37*, 126–140. [CrossRef]
24. Tyler, R.; Kaiser, W.R.; Scott, A.R.; Hamilton, D.S.; McMurry, R.G.; Zhou, N. *Geologic and Hydrologic Assessment of Natural Gas from Coal Seams in the Mesaverde Group and Fort Union Formation, Great Green River Basin, Wyoming and Colorado*; Bu-reau of Economic Geology: Texas City, TX, USA, 1994.
25. Olson, T.; Hobbs, B.; Brooks, R.; Gale, B. Paying off for tom brown in white river dom field's tight sandstone, deep coals. *Am. Oil Gas Rep.* **2002**, *10*, 67–75.
26. Li, J.S.; Li, Z.X.; Zu, B.H. Coalbed Methane (CBM) Project Enrichment Area and Economic Evaluation. *Adv. Mater. Res.* **2013**, *772*, 776–780. [CrossRef]
27. Leshchyshyn, T.T.; Rieb, B.A.; Thomson, J.T. The production suc-cess of proppant stimulation on Horseshoe Canyon coal bed methane and sandstone commingled wells. In Proceedings of the Canadian Interna-Tional Petroleum Conference, Calgary, AB, Canada, 7–9 June 2005.
28. Li, J.; Zhang, D.; Li, D.; Tang, S.; Zhang, S. Co-accumulating mechanisms of unconventional gas in the coal measure of the Qinshui Basin. *J. China Coal Soc.* **2018**, *43*, 1533–1546.
29. Cao, D.Y.; Yao, Z.; Li, J. Evaluation Status and Development Trend of Unconventional Gas in Coal Measure. *Coal Sci. Technol.* **2014**, *42*, 89–92.
30. Zhong, J.H.; Liu, C.; Wu, J.G.; Zhang, S.R.; Yang, G.Q. Symbiotic accumulation characteristics of coal measure gas in Linxing Block, eastern Ordos Basin. *J. China Coal Soc.* **2018**, *43*, 1517–1525.
31. Yi, T.S.; Zhou, X.Z.; Jin, J. Reservoir forming characteristics and co-exploration and concurrent production technology of Longtan coal measure coalbed methane & tight gas in Songhe field, western Guizhou. *J. China Coal Soc.* **2016**, *41*, 212–220.
32. Zhou, Z.; Ren, S.; Bao, S.; Chen, K.; Guo, T.; Xu, Q. The discovery of tight gas and shale gas in Silurian strata in Jianshi, Hubei Province. *Geol. China* **2018**, *45*, 855–856.
33. Lin, T.; Jin, C.S.; Wang, S.X.; Chen, K.; Pang, F.; Wang, P. A discovery of tight sandstone gas in relict synclinal of southeastern Chongqing. *Geol. China* **2017**, *44*, 814–815.
34. *SY/T 5124-2012*; Method of Determining Microscopically the Reflctance of Vitrinite in Sedimentary. Petroleum Geology Exploration Standardization Committee: Beijing, China, 2012. (In Chinese)
35. *SY/T 5125-2014*; Method of Identification Microscopically the Macerals of Kerogen and Indivision the Kerogen Type by Transmitted-Light and Fluorescence. Petroleum Geology Exploration Standardization Committee: Beijing, China, 2014. (In Chinese)
36. *GB/T 18602-2012*; Rock Pyrolysis Analysis. Chinese Standard: Beijing, China, 2012. (In Chinese)
37. *SY/T 5163-2010*; Analysis Method for Clay Minerals and Ordinary Non-Clay Minerals in Sedimentary Rocks by the X-ray Diffraction. Petroleum Geology Exploration Standardization Committee: Beijing, China, 2010. (In Chinese)

38. *GB/T18295-2001;* Analysis Method of Sandstone Sample of Petroleum and Gas Reservoir by Scanning Electron Microscope. Chinese Standard: Beijing, China, 2001. (In Chinese)
39. *GB/T 17359-2012;* Microbeam Analysis—Quantitative Analysis Using Energy Dispersive Spectrometry. Chinese Standard: Beijing, China, 2012. (In Chinese)
40. *GB/T 29172-2012;* Practices for Core Analysis. Chinese Standard: Beijing, China, 2012. (In Chinese)
41. *GB/T 17412.1~3-1998;* Classification and Nomenclature Schemes of the Rocks Classification and Nomenclature Schemes of Igneous Rock. Chinese Standard: Beijing, China, 1998. (In Chinese)
42. *GB/T 21650.1-2008;* Pore Size Distribution and Porosity of Solid Materials by Mercury Porosimetry and Gas Adsorption-Part 1:Mercury Porosimetry. Chinese Standard: Beijing, China, 2008. (In Chinese)
43. Yan, D.; Wang, H.; Fu, Q.; Chen, Z.; He, J.; Gao, Z. Geochemical characteristics in the Longmaxi Formation(Early Silurian) of South China: Implication for organic matter accumulation. *Mar. Pet. Geol.* **2015**, *65*, 290–301. [CrossRef]
44. Zou, C.N.; Dong, D.Z.; Wang, Y.M.; Li, X.; Huang, J.; Wang, S.; Guan, Q.; Zhang, C.; Wang, H.; Liu, H.; et al. Shale gas in China: Characteristics, challenges and prospects (I). *Pet. Explo-Ration Dev.* **2015**, *42*, 753–767. [CrossRef]
45. Liang, D.G.; Guo, T.L.; Bian, L.Z.; Zhao, J.; Zhao, Z. Some Progresses on Studies of Hydrocarbon Generation and Accumulation in Marine Sedimentary Regions, Southern China (Part 3): Controlling Factors on the Sedimentary Facies and Development of Palaeozoic Marine Source Rocks. *Mar. Orig. Pet. Geol.* **2009**, *14*, 1–19.
46. Wang, S.; Jiao, W.; Fang, G.; Wang, F.; Zhang, Z.; Wang, Q. Geochemical features and genesis of shale gas of Wufeng-Longmaxi Formation in Southeastern Chongqing. *Mar. Orig. Pet. Geol.* **2017**, *22*, 77–84.
47. Zhang, J.C.; Nie, H.K.; Xu, B.; Jiang, S.L.; Zhang, P.X.; Wang, Z.Y. Geological condition of shale gas accumulation in Sichuan Baisin. *Nat. Gas Ind.* **2008**, *172*, 151–156.
48. Nie, H.K.; Zhang, J.C. Shale gas accumulation condition and gas content calculation: A case study of Sichuan Basin and its periphery in the Lower Paleozoic. *Acta Geol. Sin.* **2012**, *86*, 349–361.
49. Tang, Y.; Zhang, J.C.; Liu, Z.J.; Li, L.Z. Use and improvement of the desorption method in shale gas content tests. *Nat. Gas Ind.* **2011**, *31*, 108–112.
50. Guo, T.L.; Zhang, H.R. Formation and enrichment mode of Jiaoshiba shale gas field, Sichuan Basin. *Pet. Explor. Dev.* **2014**, *41*, 28–36. [CrossRef]
51. An, Y.Y.; Yang, Z.Q.; Xiang, K.P.; Huang, Y.; Chen, H.G.; He, Y.Z.; Liu, W.; Yu, Q. The geochemical characteristics and significance of the Ordovician—Silurian black shale in northern Guizhou—A case study of the Well Daoye-1 in Daozhen County. *Geol. Rev.* **2021**, *67*, 1105–1118.
52. Zhu, Z.J.; Chen, H.D. An analysis of sedimentary characteristics and model of Silurian Xiaoheba Formation in southeastern Sichuan Province. *Geol. China* **2012**, *39*, 64–76.
53. Qiao, D.; Shi, Z.Q.; Zhang, X.; Li, C. Characteristics of contourites in the Lower Silurian in southeastern Sichuan Basin, China. *J. Palaeogeogr.* **2023**, *25*, 856–871.
54. Woodworth, J.B. On the fracture system of joints with remarks on certain great fractures. *Proc. Boston Soc. Nat. Hist.* **1896**, *27*, 163–183.
55. Murray, G.H., Jr. Quantitative fracture study: Sanish Pool, Mckenzie County, North Dakota. *AAPG Bull.* **1968**, *52*, 57–65.
56. Curtis, J.B. Fractured Shale-Gas Systems. *AAPG Bull.* **2002**, *86*, 1921–1938.
57. Gale, J.F.; Laubach, S.E.; Olson, J.E.; Eichhubl, P.; Fall, A. Natural fractures in shale: A review and new observations. *AAPG Bull.* **2014**, *98*, 2165–2216. [CrossRef]
58. Ma, S.J.; Zeng, L.B.; Shi, X.W.; Wu, W.; Tian, H.; Xue, M.; Luo, L. Characteristics and Main Controlling Factors of Natural Fractures in Marine Shale in Luzhou Area, Sichuan Basin. *Earth Sci.* **2023**, *48*, 2630–2642.

energies

MDPI

Article

Investigation of the Pore Characteristics and Capillary Forces in Shale before and after Reaction with Supercritical CO_2 and Slickwater

Chi Zhang [1,2], Qian Li [1,2,3,*], Yanlin Liu [1], Jiren Tang [1,2,3], Yunzhong Jia [1,2,3] and Tianyi Gong [1]

1 State Key Laboratory of Coal Mine Disaster Dynamics and Control, Chongqing University, Chongqing 400044, China; zhangch1.jhyt@sinopec.com (C.Z.); 202120021072t@stu.cqu.edu.cn (Y.L.); jrtang2010@163.com (J.T.); yunzhong.jia@cqu.edu.cn (Y.J.); 202220021022@stu.cqu.edu.cn (T.G.)
2 School of Resource and Safety Engineering, Chongqing University, Chongqing 400044, China
3 State and Local Joint Engineering Laboratory of Methane Drainage in Complex Coal Gas Seam, Chongqing University, Chongqing 400044, China
* Correspondence: liqian@cqu.edu.cn

Abstract: CO_2–slickwater hybrid fracturing technology is an essential part of shale gas recovery and CO_2 geo-storage. However, the exposure to supercritical CO_2 ($ScCO_2$) and slickwater can result in potential changes of the pore structures and surface wetting behavior, which affect the gas transportation and CO_2 sequestration security in shale reservoirs. Therefore, in this paper, X-ray diffraction (XRD), low-pressure nitrogen gas adsorption (N_2GA), mercury intrusion porosimetry (MIP), and fractal analysis were used to describe the pore characteristics of shale before and after $ScCO_2$–slickwater coupling treatments. Shale's surface wettability was confirmed by contact angle measurements. After the $ScCO_2$–slickwater treatments, the number of micropores (<3.5 nm) and mesopores (3.5–50 nm) increased, while that of macropores (>50 nm) declined based on the N_2GA and MIP experiments. Combined with fractal analysis, we argue that the pore connectivity diminished and the pore structure became more complicated. By analyzing the results of XRD, shale pore changes occurring after the $ScCO_2$–slickwater treatment can be explained by the adsorption of polyacrylamide (PAM). Contact angle measurement results showed that the shale's surface treated by $ScCO_2$ and slickwater was more hydrophilic than that treated by $ScCO_2$ and water, and indirectly prove our argument above. Hence, the coupling using effect of $ScCO_2$ and slickwater can impair the negative effect of CO_2 on the shale capillary force to improve shale gas productivity, but it can negatively affect the security of CO_2 sequestration in shale reservoirs.

Keywords: shale; pore structure; supercritical carbon dioxide; slickwater; CO_2 sequestration

1. Introduction

Shale gas, an unconventional natural gas resource, possesses immense development potential and vast reserves worldwide [1], making it a highly coveted commodity. The United States stands as the pioneering nation in realizing the commercial exploitation of shale gas, and its triumphant development has propelled the country towards energy autonomy. The production of U.S. shale gas has surpassed 920.65 billion cubic meters, representing a significant 74.23% of the country's overall natural gas output [2]. According to the Energy Information Administration report, China has the world's most abundant shale gas reserves, up to 31.58 trillion cubic meters, about 14.7% of the world's shale gas reserves [3]. With the increasing reliance on imported oil and gas resources, the potential for shale gas development is also escalating, thereby rendering the exploration of shale gas a highly sought-after research topic.

Shale reservoirs are typically remarkably tight, with low porosity and permeability, and a large amount of shale gas exists in an adsorbed state, so shale reservoirs need to be stimulated to increase their permeability before shale gas can be extracted [4]. Horizontal well hydraulic fracturing serves as the primary technology for extracting shale gas. Nevertheless, this process is accompanied by numerous drawbacks, including significant water resource consumption, potential contamination of underground water, and the challenge in reclaiming and draining fracturing fluids [5]. Hence, supercritical carbon dioxide has emerged as a promising alternative to traditional fracturing fluids for the purpose of hydraulic fracturing. This specialized phase of carbon dioxide, situated between liquid and gas, offers unique advantages for this application. Its temperature and pressure are higher than the critical temperature and pressure (Tc = 31.04 °C, Pc = 7.38 MPa) [6]. It has a number of unique physical and chemical properties, such as a low viscosity, a high diffusion coefficient, a surface tension close to zero, etc., [6]. Capable of penetrating the micropores of shale reservoirs to generate micro-cracks during fracturing, it is deemed an optimal substitute for water-based fracturing fluids in hydraulic fracturing operations [7]. However, it has been observed that the proppant-carrying capacity of supercritical carbon dioxide is limited due to its extremely low viscosity, resulting in the inability to generate effective fractures post-fracturing [8]. Furthermore, the exclusive utilization of CO_2 for hydraulic fracturing fails to create primary fractures of significant length and width within the geological formations, thereby limiting the scope of the fracture stimulation [9]. Therefore, a hybrid fracturing technique has been proposed, which utilizes the combination of supercritical CO_2 and slickwater. This approach initially involves injecting liquid CO_2 into the shale formation for fracturing purposes, during which the CO_2 transitions to a supercritical state [9]. This approach can significantly augment the intricacy and extensive development of the fracture network within the shale reservoir, thereby expediting reservoir activation and minimizing the fracture initiation pressure [10]. Subsequently, slickwater with proppants is injected at high pressure to fracture the shale reservoir, effectively supporting the fractures and creating a large-scale fracture network to enhance the stimulated volume of the reservoir [11]. Based on field application outcomes, the hybrid fracturing technology utilizing supercritical carbon dioxide along with slickwater has demonstrated its effectiveness in bolstering the extraction efficiency of shale gas [12]. Furthermore, shale exhibits a greater affinity for CO_2 compared to CH_4, indicating that the injection of CO_2 can efficiently displace CH_4, ultimately leading to a significant enhancement in shale gas recovery [13].

Moreover, utilizing CO_2 to enhance shale gas development effectively serves the dual purpose of advancing both shale gas extraction and CO_2 geological sequestration. Notably, CO_2 is a significant greenhouse gas that contributes significantly to global warming. Various countries have recently developed carbon neutrality roadmaps to mitigate climate change. According to the 'China Carbon Dioxide Capture, Utilization, and Storage (CCUS) Annual Report (2021)', the global onshore potential for CO_2 geological sequestration is estimated to be between 6×10^{12} and 42×10^{12} tons [14]. This is 165 to 1157 times the global CO_2 emissions from energy combustion and industrial processes in 2021, which totaled 363×10^{12} tons [14]. This underscores the immense potential of CO_2 geological sequestration as a pivotal tool in attaining carbon neutrality [15]. Currently, the primary geological formations used for CO_2 sequestration include oil and gas reservoirs, deep saline aquifers, and unmineable coal seams [15]. Oil and gas reservoirs, traditionally utilized for hydrocarbon storage, offer unwavering seal integrity. Furthermore, CO_2 can effectively boost the recovery of oil and gas resources, thus rendering its technical and economic sequestration more viable. Thanks to their extensive fracture networks, depleted shale gas reservoirs provide a significant storage capacity for CO_2 [16]. The unfractured shale layers also possess the functionality of cap rocks, effectively preventing the leakage of CO_2. As a result, these depleted reservoirs are deemed to be highly suitable geological locations for the sequestration of CO_2.

During the initial phase of shale gas extraction and the subsequent CO_2 geological sequestration, key properties such as pore characteristics and wettability hold significant importance [17]. The pore characteristics serve as an indicator of the number of channels present within the shale formation, whereas wettability encapsulates the mechanical interactions between the solid and liquid phases within the reservoir [17]. Collectively, these factors exert a considerable impact on both the extraction efficiency of shale gas and the stability of CO_2 geological sequestration [18]. However, after the hybrid fracturing process utilizing CO_2 and slickwater, there is an inevitable impact on the shale reservoir due to the residual CO_2 and slickwater [9]. For instance, the carbonic acid that results from the reaction between CO_2 and slickwater has the ability to interact with the mineral components present in shale [9]. Supercritical CO_2 may also dissolve organic materials and clay minerals [19]. These interactions have the potential to modify the pore structure and wettability of shale, which in turn can significantly influence both the recovery of shale gas and the effectiveness of CO_2 geological sequestration [20].

The ways in which individual fracturing fluids interact with shale and the resulting changes in the shale pore structure are largely understood. However, despite the prevalent application of combined fracturing techniques, there remains a scarcity of research investigating the combined impact of $ScCO_2$ and slickwater on the pore structure and wetting behavior of terrestrial shale formations. This gap in knowledge hinders the accurate evaluation of pore characteristics and wettability in the formation after fracturing. Therefore, it is essential to conduct exposure experiments on shale saturated with slickwater under different temperatures and CO_2 pressures. Subsequent analyses of the treated samples using techniques such as X-ray diffraction (XRD), atomic force microscopy (AFM), low-temperature N_2 adsorption, mercury intrusion porosimetry (MIP), and contact angle measurements can be performed to characterize the shale samples. These tests aim to explore the variations in shale mineral components, surface morphology, pore characteristics, and wettability. This research will aid in enhancing CO_2-enhancing shale gas development and CO_2 geological storage technologies.

2. Materials and Methods

2.1. Sample Description and Preparation

The samples for the experimental study were collected from the terrestrial shale of Zheng 083 well in the Chang 7 section of the Upper Triassic Extension Formation in the Zhangjiatan area, Ordos Basin. The samples were buried at a depth of 495 m~512 m. Upon completion of coring, the cores were promptly wrapped in parafilm and dispatched to the laboratory to minimize water evaporation and oxidation of the samples due to air exposure [9].

In this study, 1.5% KCl (as an anti-expansion agent) was used as the base liquid, and polyacrylamide (PAM, drag-reducing agent) with a concentration of 0.1% was blended with the slickwater. After mixing and blending the slickwater, it was left for 4 h to stabilize its viscosity. The resistivity of the distilled water used was greater than 18 MΩ·cm, and the KCl powder was analytically pure. PAM is a water-soluble anionic polymer with a molecular weight of 16 million. The viscosity of the prepared slickwater was measured to be 5.7 mPa·s and 4.7 mPa·s at room temperature and 40 °C, respectively, and was analyzed using an Anton Paar rheometer (model MCR302, Graz, Austria).

2.2. CO_2 Treatment Procedure

The soaking equipment used in this experiment was a self-designed high-temperature and high-pressure shale soaking device, which includes a water bath with an accuracy of ±0.05 °C, a high-temperature and high-pressure reactor with internal dimension φ60 mm × 120 mm, a vacuum pump, a syringe pump (Teledyne ISCO 260D) and a pressure sensor. The reactor's maximum working pressure and temperature are up to 40 MPa and 100 °C, respectively [21].

In order to analyze and explain the mechanism of how to change the pore structure and capillary pressure of shale when treated with $ScCO_2$–slickwater, $ScCO_2$-distilled water treatment and He-slickwater treatment were added as a control group in the experiment. The experiment was divided into three steps. Step 1: Shale saturated with distilled water/slickwater. The shale bulk was cracked into particles with a grain size of 10–20 mm and put into a water saturator with a pressure of 0.1 MPa and saturated with distilled water/slickwater until the pressure no longer changed and then taken out. Step 2: $ScCO_2$ static soaking experiments of saturated distilled water/slickwater shale were carried out using the above-mentioned soaking device. To simulate the effect of $ScCO_2$ on the pore structure of shale after entering the formation, the shale was treated at 40 °C and 8 MPa for 4 days. Then, shale particles were dried in a vacuum oven at 110 °C for 48 h to completely remove the free and weakly bound water in the shale pores. Step 3: According to the requirements of different testing approaches, the treated shale was crushed into powders of different grain sizes for testing and analysis. XRD was used to characterize the changes in the mineral compositions of the shale, combined with N_2 adsorption to characterize the changes in the pore parameters of the shale before and after treatment.

2.3. XRD Analysis

Before XRD analysis, the samples were ground to less than 200 mesh (75 μm). Powder samples were tested at a rate of 2°/min ranging from 3° to 45° by an X-ray diffractometer (Bruker D8 Advance, Karlsruhe, Germany) to determine the relative mineral contents of shale before and after pure CO_2/CO_2 + slickwater treatments. The relative mineral content was obtained by fitting the XRD spectra using Jade 6 software [22].

2.4. N_2 Adsorption Analysis

Powder samples were tested by a surface area and pore size distribution analyzer (Belsorp-Max II, Matsumoto, Japan) to determine the pore structure of the shale before and after pure CO_2/CO_2 + slickwater treatments. Prior to testing, the shale underwent crushing into 60–80 mesh particles, aiming to expedite the adsorption equilibrium time while simultaneously enhancing the effectiveness of pore testing. In order to maximize the removal of residual gases and moisture in the pores without destroying the pore structure and to improve the accuracy of the test, the shale was degassed under vacuum at 110 °C for 15 h [23]. Then, the shale particles were subjected to N_2 adsorption experiments at −196 °C and the relative pressures (P/P_0) of 0.01~0.99 to obtain the N_2 adsorption–desorption isotherms. Specific surface area (SSA) was calculated by the Brunauer–Emmet–Teller (BET) model [24]. The pore volume and surface area of meso- and macro-pores were evaluated by the application of the BJH model that accounts for the volume desorbed via both capillary evaporation (described by the Kelvin equation) and thinning of multilayers of adsorbed molecules in cylindrical pores. In order to eliminate the artificial error caused by the tensile strength effect, the adsorption branch was selected for the BJH model in this paper. The surface area and pore volume of micropores were estimated by the HK model and the total pore volume (TPV) was set as the liquid N_2 volume at the highest P/P_0. The average pore diameter (Da) was calculated from $4TPV/S_{BET}$ and the pore size distribution was derived from the nonlocal density function theory (NLDFT).

2.5. Mercury Intrusion Porosimetry (MIP) Test

A mercury intrusion pore size analyzer (PoreMaster-33, Quantachrome, Boynton Beac, FL, USA) was used to perform mercury intrusion experiments up to 33,000 psi (227 MPa), which corresponds to a minimum pore diameter of 6.4 nm according to the Washburn Equation (1). The shale samples used to conduct the MIP tests were crushed into 1–3 mm particles and then degassed at 110 °C for 15 h before being subjected to MIP testing [25] as follows:

$$P_m = -\frac{4\sigma cos\theta}{d} \tag{1}$$

where P_m is the capillary pressure, MPa; σ is the surface tension of mercury, 0.48 N/m; θ is the contact angle between shale and mercury, 130°; and d is the pore diameter, μm. In general, the N_2 adsorption method is mainly used to characterize micropores and mesopores, while the MIP method can better characterize macropores and make up for the shortcomings of the N_2 adsorption method; so, the combination of the two methods can characterize the pore structure of shale before and after treatment in a more complete way.

2.6. Contact Angle Measurement

The shale for the contact angle experiments was prepared as thin sections (about 10 mm × 10 mm × 2 mm), the surface of the shale samples was polished carefully with 1200 mesh sandpaper, and cleaned and dried to minimize the effect of surface roughness on the contact angle. First, the roughness analysis of the shale surface before and after treatment was carried out using an atomic force microscope (AFM) [17]. Then, the contact angle test was carried out at room temperature using a JY-PHa (Sichuanjinhe Co., Mianyang, China) contact angle tester. Five measuring points were selected for each sample, the drop volume of liquid was 12 μL, and the contact time was 60 s [26].

3. Results

3.1. Mineral Compositions before and after Treatment

The mineral compositions of the shale samples before and after treatment were determined by quantitative and qualitative analysis of XRD patterns (Figures 1 and 2 and Table 1). Untreated shale is rich in quartz (SiO_2) and clay minerals (kaolinite, illite and clinochlore), of which the content of quartz is 24.2% and the total clay minerals is 48.89%. Due to the significant clay mineral content, the interaction between water and clay within shale results in the swelling of the clay during the water-based fracturing processes. The composition of plagioclase stands at 14.27%, while calcite ($CaCO_3$) comprises 9.43%. On the other hand, dolomite ($Ca\,Mg\,(CO_3)_2$) and pyrite (FeS_2) account for 1.09% and 1.12%, respectively, both of which are under 3% in concentration. After undergoing treatment with $ScCO_2$–slickwater, the quartz content underwent a reduction of 9.13%. Additionally, the calcite and clay mineral contents exhibited specific alterations. Specifically, the kaolinite content increased by 3.94%, while the illite content rose by 14.2%. Conversely, the clinochlore content decreased by 9.31%. Following the treatment, the dolomite content exhibited a certain degree of increase, whereas the pyrite content declined accordingly, albeit with minimal overall variation. Following treatment with $ScCO_2$–distilled water, the quartz content decreased by 1.38%, while the calcite content saw a reduction of 0.77%. Furthermore, the overall clay mineral content declined by 0.24%, specifically due to reductions in the kaolinite and clinochlorite contents, with a corresponding increase in the illite content. After undergoing treatment with He–slickwater, the quartz content underwent a slight reduction of 3.91%, whereas the calcite and clay mineral contents (inclusive of kaolinite, illite, and clinochlorite, all experiencing an upsurge) increased by 2.93% and 6.24%, respectively. These mineral changes may be related to complex chemical reactions among shale, CO_2, and slickwater [9]. The dissolution of CO_2 in water leads to the formation of carbonic acid, and under weakly acidic conditions, a series of reactions are prone to occur in shale minerals [26]:

$$CO_2 + H_2O \leftrightarrow H_2CO_3 \leftrightarrow H^+ + HCO_3^- + CO_3^{2-} \tag{2}$$

$$\text{calcite} + 2H^+ \leftrightarrow Ca^{2+} + CO_2 + H_2O \tag{3}$$

$$\text{dolomite} + 4H^+ \leftrightarrow Ca^{2+} + Mg^{2+} + 2CO_2 + 2H_2O \tag{4}$$

$$\text{illite} + 1.1H^+ \leftrightarrow 0.77\text{kaolinite} + 0.6K^+ + 0.25Mg^{2+} + 1.2\text{quartz} + 1.35H_2O \tag{5}$$

$$\text{pyrite} + O_2(aq) + 4H^+ \leftrightarrow 4Fe^{3+} + 2H_2O \tag{6}$$

$$\text{plagioclase} + 8H^+ \leftrightarrow Ca^{2+} + 2Al^{3+} + 2SiO_2 + 4H_2O \tag{7}$$

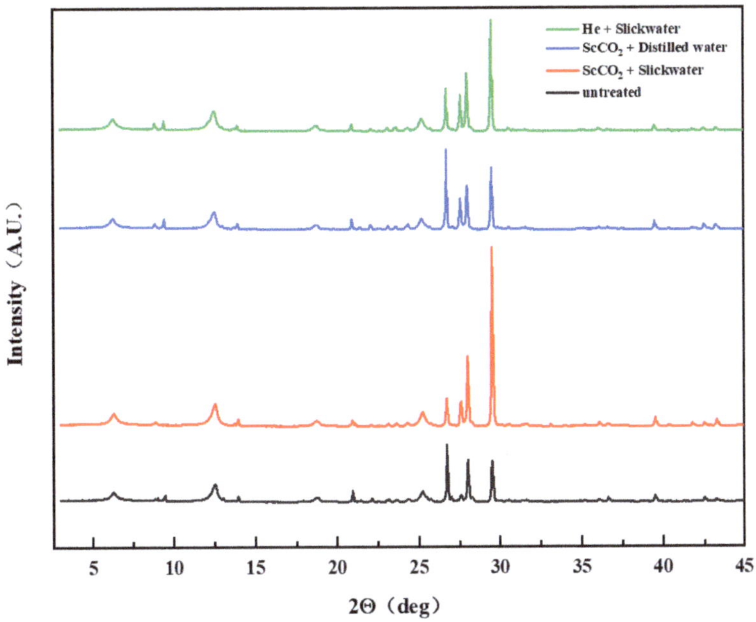

Figure 1. The XRD of untreated sample (black), $ScCO_2$ + slickwater-treated sample (red), $ScCO_2$ + distilled water-treated sample (blue), and He + slickwater-treated sample (green).

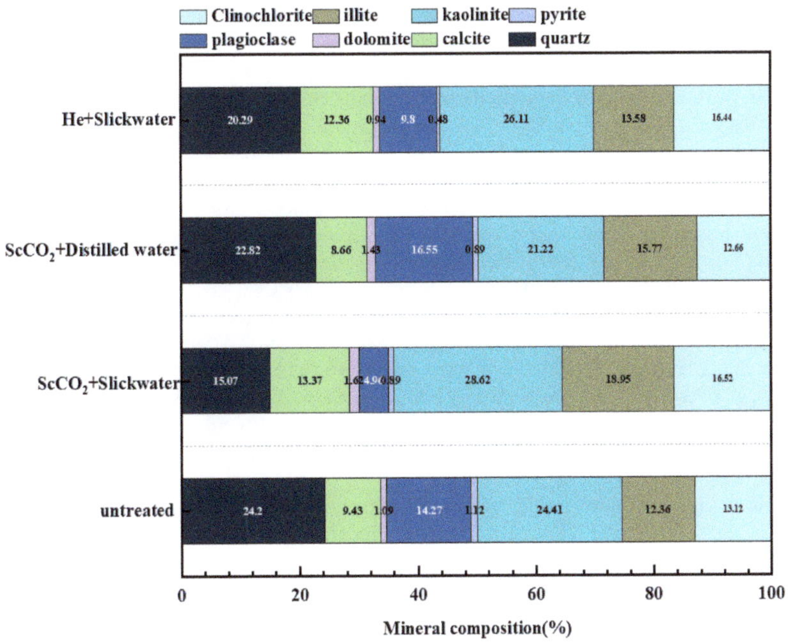

Figure 2. The mineral contents of the untreated sample, $ScCO_2$ + slickwater-treated sample, $ScCO_2$ + distilled water-treated sample, and He + slickwater-treated sample were obtained by XRD analysis.

Table 1. Mineral composition of the shale samples before and after treatment.

Sample	Quartz	Calcite	Dolomite	Plagioclase	Pyrite	Total Clay	Kaolinite	Illite	Clinochlore
Untreated	24.2	9.43	1.09	14.27	1.12	49.89	24.41	12.36	13.12
$ScCO_2$–slickwater	15.07	13.37	1.62	4.96	0.89	64.09	28.62	18.95	16.52
$ScCO_2$–distilled water	22.82	8.66	1.43	16.55	0.89	49.65	21.22	15.77	12.66
He–slickwater	20.29	12.36	0.94	9.8	0.48	56.13	26.11	13.58	16.44

As the duration of soaking increases, CO_2 undergoes slow solubility in water, leading to the production of carbonic acid [27]. This process results in a decrease in the pH level of the system, thereby enhancing the dissolution and precipitation reactions of minerals. Multiple studies have unequivocally demonstrated that calcite possesses the utmost solubility within the CO_2–water–shale system, while minerals akin to feldspar exhibit a similar trend of solubility [28]. Their solubility mostly influences the process of mineral dissolution.

Table 1 indicates that following $ScCO_2$–slickwater treatment, the quartz and clinochlorite content underwent a significant reduction of 37.7% and 65.2%, respectively, while the calcite and clay minerals exhibited a marked increase of 41.78% and 28.46%, respectively. According to Equations (3), (4), and (7) above, the dissolution of plagioclase leads to an increase in the concentration of Ca^{2+}, SiO_2, and the pH value in the $ScCO_2$–slickwater system. SiO_2 is the main component of quartz, so it may lead to a reverse reaction in Equations (3) and (4), increasing the content of the calcite, dolomite, and clay minerals. According to Equation (5), an illite reaction with $ScCO_2$–slickwater consumes hydrogen ions, raises the pH, and produces magnesium ions and quartz. According to Equation (7), the reaction of plagioclase with $ScCO_2$–slickwater also consumes H^+, raises the pH, and produces Ca^{2+} and quartz. On the one hand, the aforementioned reactions supply Ca^{2+} and M^{2+}; on the other hand, they raise the pH, prompting Ca^{2+} and M^{2+} to react with CO_2, resulting in the formation and precipitation of the carbonate minerals calcite and dolomite. This explains why, following $ScCO_2$–slickwater treatment, the calcite and dolomite contents of the samples underwent a significant increase, while the plagioclase content underwent a notable decrease [29].

Interestingly, samples treated with slickwater exhibited a significant increase in the clay content, in stark contrast to those treated with distilled water, which displayed virtually no alteration in the clay content. This discrepancy may be attributed to the substantial presence of K^+ in slickwater. Furthermore, following the $ScCO_2$–distilled water and He–slickwater treatments, the quartz content exhibited respective declines of 5.7% and 16.1%, whereas other mineral components underwent varied alterations. After undergoing $ScCO_2$–distilled water treatment, the calcite and kaolinite content underwent a reduction of 8.2% and 13.1%, respectively, while the total clay mineral content decreased by 0.5%. Conversely, following treatment with He–slickwater, the calcite and kaolinite content saw an increase of 31.7% and 7.0%, respectively, leading to an overall 12.5% increase in the total clay mineral content.

Equation (6) demonstrates that the process of pyrite oxidation in shale is regulated by the level of dissolved oxygen [30]. During the experimental procedure, the utilization of deionized water for slickwater preparation, the saturation of shale under vacuum conditions, and the implementation of a CO_2 purification unit prior to the immersion of $ScCO_2$ led to restricted levels of ambient and dissolved oxygen. Moreover, the $ScCO_2$–slickwater system demonstrates a minimal proportion of a pyrite oxidation reaction, attributable to the low initial concentration of pyrite. This phenomenon can be attributed to the sluggish oxidation reaction occurring within the system, resulting in a diminished pyrite content within the shale.

In summary, the XRD results of the controlled experiment reveal that the decrease in the shale calcite content upon treatment with $ScCO_2$–distilled water is notably less significant than the increase in the shale kaolinite content following treatment with $ScCO_2$–slickwater. This phenomenon can be attributed to the fracturing fluid, such as slickwater, which enhances the network pore structure and facilitates increased interaction between CO_2 and minerals [31]. Consequently, water exhibits more mineral reactions than distilled water [9]. This observation suggests that the mineral mechanism reaction in shale may be more intricate under real reservoir conditions. Shale treated with $ScCO_2$–slickwater has a higher solubility of plagioclase than the He–slickwater treatment. Although He is insoluble in water, the introduction of $ScCO_2$ enhances the acidity of the slickwater system, subsequently leading to a decrease in the dissolution of plagioclase following treatment with $ScCO_2$–slickwater [9]. In contrast, it is observed that the kinetic diameter of He molecules (0.28 nm) is comparatively smaller than that of CO_2 molecules (0.33 nm). This advantage allows the slickwater to penetrate a reduced number of apertures and achieve effective engagement with the shale. This study's findings reveal that the concentration of quartz in the He–slickwater treatment surpassed that observed in the $ScCO_2$–slickwater treatment. Based on the observed changes in the mineral composition, we can deduce that the shale reaction extent following the $ScCO_2$–water treatment is anticipated to be within the spectrum of both the $ScCO_2$–water treatment and sole slickwater treatment. This observation lays the groundwork for anticipating the geochemical interplay between fluids and solids, as well as the transformation of the pore structure within the formation following fracturing.

3.2. Analysis of the Pore Characteristics Based on N_2 Adsorption Isotherms

3.2.1. Analysis of the Pore Characteristics Based on N_2

Figure 3 displays the low-pressure N_2 adsorption–desorption isotherms of the samples. The adsorption isotherms of the samples are classified as type IV(a) by the International Union of Pure and Applied Chemistry (IUPAC) [32]. Additionally, the hysteresis loop of the samples is identified as type H3. This classification signifies that the shale pores are predominantly comprised of slit pores, interspersed with a minority of pores exhibiting the characteristic inkbottle-shaped pore throats. The samples' adsorption–desorption isotherms and hysteresis loops exhibited a comparable morphology before and after treatment, suggesting that the pore morphology remained rather stable [33]. The isotherms may be categorized into three stages based on the relative pressure (P/P_0). When the ratio of P/P_0 is less than 0.45, the adsorption curve gradually rises as the pressure increases. This observation suggests that N_2 effectively occupies the shale's micropores, whereas single-layer adsorption occurs in the medium and large pores [34]. When the ratio of P/P_0 exceeds 0.45, the slope of the curve exhibits an upward trend as the ratio of P/P_0 gradually increases [34]. This observation indicates that N_2 is adsorbed onto the shale surface in multiple layers, resulting in the phenomenon of capillary condensation.

When the P/P_0 is more than 0.9, the curve and gas adsorption capacity experience a significant rise as a result of capillary condensation in the macropores [9]. As the relative pressure nears 1, N_2 adsorption remains unsaturated, suggesting that a considerable percentage of macropores possess diameters exceeding the scope of the N_2 adsorption technique. Therefore, mercury intrusion porosimetry serves as an indispensable complementary instrument. The untreated shale sample exhibits a larger adsorption capacity than the shale treated with $ScCO_2$–slickwater, $ScCO_2$–distilled water, and He–slickwater at the highest relative pressure (P/P_0 approaches 1). This observation suggests that the adsorption capacity of the shale undergoes alterations during the treatment process. Prior research has indicated that the adsorption capacity of shale is primarily influenced by micropores and mesoporous, with a particular emphasis on micropores [35]. Hence, an analysis was conducted on the pore characteristics of each pore in the shale both prior to and after treatment.

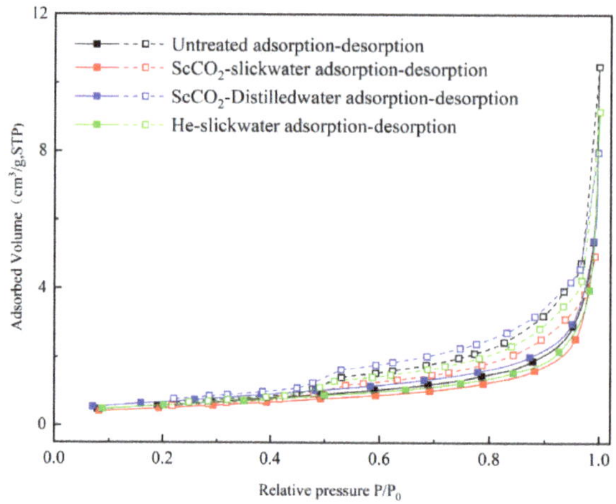

Figure 3. Adsorption–desorption isotherms of the shale before and after treatment.

3.2.2. Changes of the Pore-Structure Parameters Based on N_2

Table 2 summarizes the parameters of the pores before and after treatment based on the N_2 adsorption and desorption. As can be seen in Table 2, the contribution of micropores to the total specific surface area and total pore volume is minimal; mesopores contribute the most to the specific surface area (more than 80%), and mesopores and macropores contribute equally to the total pore volume. The results showed that the specific surface area (S_{BET}), total pore volume (TPV), and average pore diameter (Da) decreased after the $ScCO_2$–slickwater treatment. However, after the $ScCO_2$–distilled water treatment, the S_{BET} and Da decreased, but the TPV increased. In addition, the specific surface area and pore volume corresponding to the micropores, mesopores and macropores were reduced after the treatment with $ScCO_2$–slickwater. This phenomenon indicates that the $ScCO_2$–slickwater treatment can simultaneously affect micro-, meso-, and macropores.

Table 2. SSA, PV, and Ra of the shale pores based on N^2.

Sample	Surface Area (m^2/g)			Pore Volume (cm^3/kg)			S_{BET} (m^2/g)	TPV (cm^3/kg)	Da (nm)
	Micro (HK)	Meso (BJH-ad)	Macro (BJH-ad)	Micro (HK)	Meso (BJH-ad)	Macro (BJH-ad)			
Untreated	0.153	2.018	0.139	0.1299	5.384	6.374	2.128	16.199	30.443
$ScCO_2$–slickwater	0.114	1.712	0.090	0.0976	4.560	2.943	1.822	7.656	16.812
$ScCO_2$–distilled water	0.207	1.982	0.140	0.174	5.521	6.702	2.477	12.341	19.932
He–slickwater	0.116	1.813	0.144	0.100	4.549	6.222	2.081	14.175	27.247

Upon treatment with $ScCO_2$–distilled water, the specific surface area of the micropores underwent a significant augmentation, whereas the corresponding areas of the meso- and macropores exhibited negligible variation. Correspondingly, the pore volume of the micropores increased substantially, and the pore volume of the meso- and macropores increased slightly. The S_{BET} increased, while the TPV and the Da both decreased. The aforementioned findings suggest that $ScCO_2$–distilled water primarily exerts a notable influence on the microporosity characteristics of shale. The sample, following treatment with He–slickwater, exhibited a diminished specific surface area of micropores and mesopores, coupled with a marginal increase in the specific surface area of the macropores. The pore volume within the micropores, mesopores, and macropores underwent a decline. Similarly, the S_{BET} and

TPV also decreased, accompanied by a reduction in the average pore diameter. Given that helium is an inert gas, this shift in pore structure parameters is attributed to the influence of the slickwater [9].

The pore size distribution (PSD) of the shale samples before and after treatment evaluated from the NLDFT model are presented in Figure 4. As observed in the PSD curves depicted in Figure 4, it is evident that the micropores within the samples exhibit limited development. However, the mesopores, particularly those falling within the range of 3 nm to 40 nm, are more prominently developed, with the peak concentration observed in pores possessing diameters of 4–6 nm. Despite the alterations in the pore parameters, it is noteworthy that the PSD trend of the treated samples remained unchanged. For example, the PSD peaks of all the samples appeared at 4–6 nm, which is consistent with the fact that the shape of the N_2 adsorption–desorption isotherm did not change substantially. Upon treating the samples with $ScCO_2$–slickwater, it was observed that the differential pore volume distribution diminishes within the range of 2–30 nm. This finding aligns with the results presented in Table 2, indicating a reduction in the pore volume of the mesopores. After treatment with $ScCO_2$–distilled water, the samples exhibited a significant increase in the differential pore size distributions of both the micropores and mesopores. Notably, the peak increase at 4–6 nm was particularly prominent, aligning with the augmentation in the pore volume observed for both micropores and mesopores, as detailed in Table 2. For the samples treated with He–slickwater, the differential pore size distribution increases significantly at 3–5 nm while decreasing slightly at 7–30 nm.

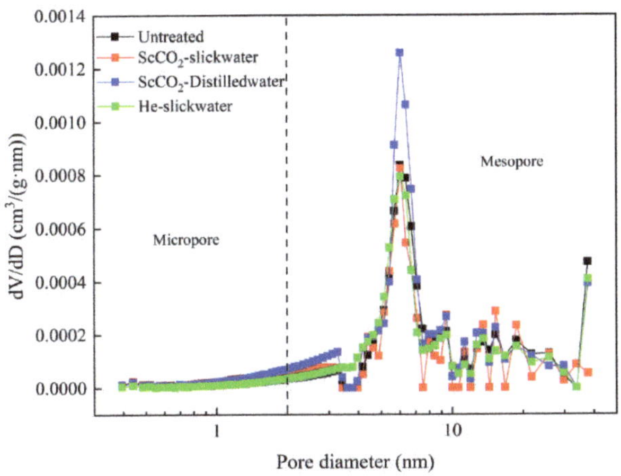

Figure 4. Pore-size distributions before and after treatment based on N_2.

3.2.3. Fractal Dimensions of Sample Based on N_2 Isotherms

The fractal dimension of the N_2 isotherm can be calculated using the widely used Frenkel–Halsey–Hill (FHH) model, which assumes that the thickness of the adsorbate layer on a pore surface is directly related to the fractal dimension of that surface [36]. Moreover, it implies that the adsorbate molecules' behavior is influenced by the potential field created by the adsorbent surface, as follows:

$$ln\frac{V}{V_m} = C + A\ln\left[ln\left(\frac{P_0}{P}\right)\right] \tag{8}$$

where V implies the volume of the adsorbed N_2, cm^3/g at STP; the value V_m is the maximum volume of N_2 needed to cover a monolayer, measured in cm^3/g at STP completely and it can be determined by using the BET model; P_0 represents the saturated vapor pressure of N_2, MPa; P denotes the equilibrium pressure of N_2 adsorption, MPa; C is a constant.

By graphing $\ln(V/V_m)$ against $\ln[\ln(P_0/P)]$ using a double logarithmic coordinate system, a linear connection may be shown. The slope of this line corresponds to the value of A.

Two types of forces predominantly govern the adsorption of gases on pore surfaces: van der Waals forces (VDW) and capillary condensation [37]. The VDW mechanism is the main process involved in the formation of a monolayer. Primarily, the interactions between the solid and vapor phases govern this process. Conversely, capillary condensation occurs in a later stage of adsorption, primarily influenced by the interactions between the liquid and vapor phases. Researchers have documented two distinct fractal dimension values, each stemming from the respective relative pressure ranges of 0–0.5 and 0.5–1.0. The fractal dimension is evaluated within the VDW force regime, spanning relative pressures from 0 to 0.05, and during the capillary condensation phase, encompassing relative pressures from 0.5 to 1.0.

$$D_2 = A + 3 \tag{9}$$

Figure 5 demonstrates that the FHH model accurately depicts the fractal characteristics for relative pressures ranging from 0 to 0.5 and from 0.5 to 1. The correlation coefficients (R^2) are equal to or greater than 0.977 for all samples, except for the untreated sample at a relative pressure >0.5, which suggests that the FHH model is applicable to N_2 adsorption on shale. The gentler incline on the left side of the FHH plot is associated with capillary condensation when the adsorbate gathers within the pores. On the other hand, the greater degree of incline on the right side of the FHH graph is interpreted as evidence of both single-layer and multiple-layer adsorption taking place within the pores [38]. When the relative pressure is less than 0.5, the fractal dimension increases gradually with the untreated sample, $ScCO_2$–slickwater-treated sample, $ScCO_2$–distilled water-treated sample, and He–slickwater-treated sample, as shown in Table 3. Extensive research studies have put forward the theory that the fractal dimension observed at relatively low pressures corresponds to the influence exerted by van der Waals (VDW) forces, thereby serving as an indicator of the surface fractal dimension. This, in turn, can effectively represent the degree of roughness present on the surface of the pores.

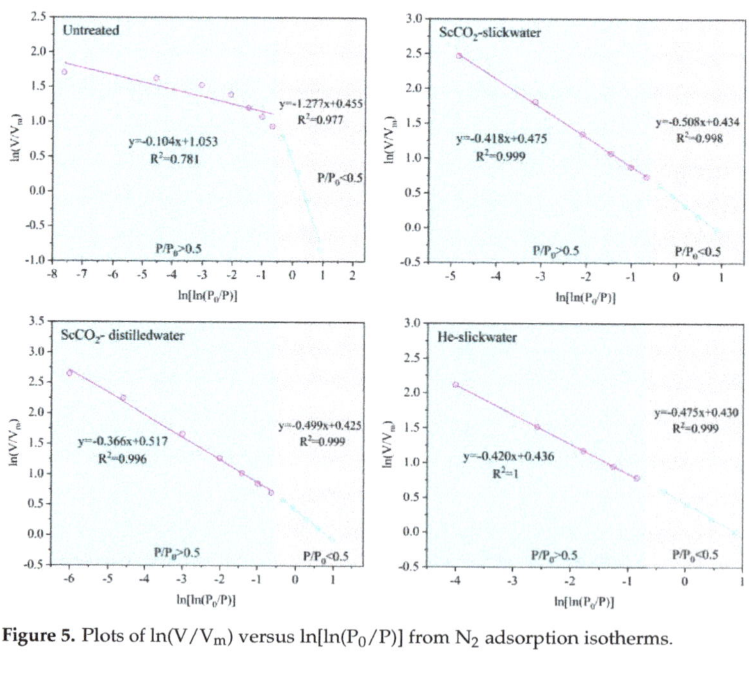

Figure 5. Plots of $\ln(V/V_m)$ versus $\ln[\ln(P_0/P)]$ from N_2 adsorption isotherms.

Table 3. The parameters of fractal dimensions of shale sample derived from the FHH model.

Sample	$P/P_0 < 0.5$			$P/P_0 > 0.5$		
	Fitting Equation	D	R^2	Fitting Equation	D	R^2
Untreated	$y = -1.277x + 0.455$	1.723	0.977	$y = -0.104x + 1.053$	2.896	0.781
$ScCO_2$–slickwater	$y = -0.508x + 0.434$	2.492	0.998	$y = -0.418x + 0.475$	2.582	0.999
$ScCO_2$–distilled water	$y = -0.499x + 0/425$	2.501	0.999	$y = -0.366x + 0.517$	2.634	0.996
He–slickwater	$y = -0.475x + 0.430$	2.525	0.999	$y = -0.420x + 0.436$	2.580	1

This indicates that the sole impact of slickwater results in significant alterations to the roughness and intricacy of the shale's micropores. This aligns with the observation that both the specific surface area and pore volume of the micropores underwent a substantial decrease in the sample subjected to the He–slickwater treatment. When slickwater interacts with $ScCO_2$ or solely $ScCO_2$ combines with distilled water, it likewise elevates the roughness and intricacy of the micropores, albeit to a lesser extent than $ScCO_2$–slickwater. The fractal dimension under elevated relative pressures corresponds to the phenomenon of capillary condensation and signifies the fractal dimension of the pore structure, which can be utilized to delineate the irregularity of pores [38]. Table 3 reveals that the $ScCO_2$–slickwater and He–slickwater treatments both significantly mitigate the irregularity of the pore structure, though the $ScCO_2$–distilled water treatment exhibits a relatively modest reduction in pore structure irregularity.

3.3. Analysis of the Pore Characteristics Based on MIP

The mercury intrusion and extrusion curves for untreated and treated shale samples are depicted in Figure 6. It is evident that the shapes of the intrusion and extrusion curves for the treated shale samples remain largely unchanged, suggesting a minimal alteration in the pore throat geometry of these samples. This observation aligns well with the variations observed in the N_2 adsorption and desorption isotherms prior to and following the treatment process. The mercury intrusion and extrusion curves can be divided into three segments. When the pressure drops below 0.1 MPa, a rapid surge in mercury content signifies the existence of macroscopic cracks within the samples. Upon the increment of pressure from 0.1 MPa to 10 MPa, a near-linear surge in the mercury quantity was observed, suggesting a notably consistent distribution of macropores. Upon an increase in intrusion pressure, from 10 MPa to 227 MPa, a notable acceleration in the mercury volume was observed, suggestive of a substantial presence of pore throats within the mesopore size range in the samples.

The pore size distribution of shale is shown in Figure 7. The untreated shale is dominated by mesopores, accounting for 90.33% of the total porosity. After treatment with $ScCO_2$–slickwater, $ScCO_2$–distilled water, and He–slickwater, the TSSA and TPV of the shale show an increasing trend (Table 4). Observing Figures 8 and 9, it is evident that the SSA and PV values within the mesopores underwent a decrease of 0.94% and 1.74%, respectively, while the macropore proportion experienced a corresponding increase. Micropores and fine mesopores mainly determine the SSA, and mesopores mainly determine the PV. Therefore, the TPV increases with the increase in the number of macropores. The MIP data showed the Da increased after the $ScCO_2$–distilled water treatment and decreased after the He–slickwater treatment. Therefore, the reasons for the pore changes caused by the two treatments are different [39]. The process of extraction and dissolution, stemming from the treatment with $ScCO_2$–distilled water, has the potential to transform smaller shale pores into larger ones, thereby causing a decrease in the TSSA as the number of mesopores increases. In addition, the precipitation of secondary minerals in the pores may lead to clogging of the macropores, resulting in a reduced TPV. The relative content of different pore types determines the Da. After the $ScCO_2$–distilled water treatment, the proportion of macropore volume increased by 2.55%, and the Da also increased correspondingly. In addition to the mineral dissolution and ion precipitation caused by the water environment, PAM adsorption is also the reason for the decreased shale pore number after the He–slickwater treatment [40]. The gelatinized PAM may enter the macropores under a certain gas pressure and adhere to the inner wall of the shale pores, thus reducing the Da. As a result, the enlargement in the pore volume within the shale subjected to the $ScCO_2$–slickwater treatment can be attributed to the extraction and dissolution of minerals. Conversely, the reduction in the macropore volume may stem from the combined influence of ion precipitation and PAM adsorption [41].

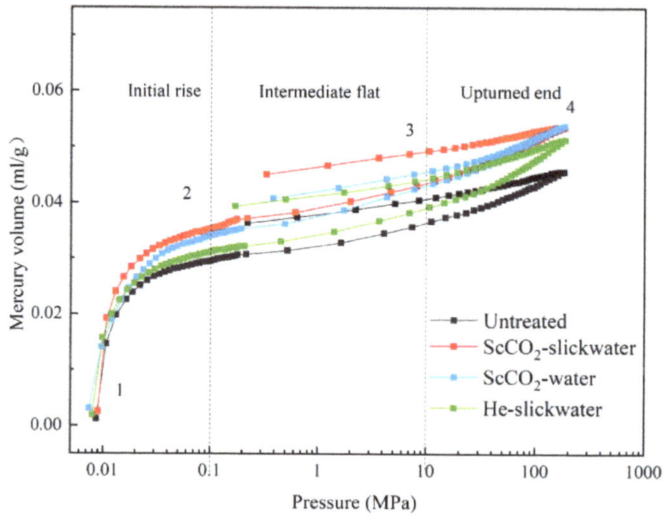

Figure 6. The intrusion and extrusion curves of mercury for the untreated sample (black), $ScCO_2$–slickwater-treated sample (red), $ScCO_2$–distilled water-treated sample (blue), and He–slickwater-treated sample (green).

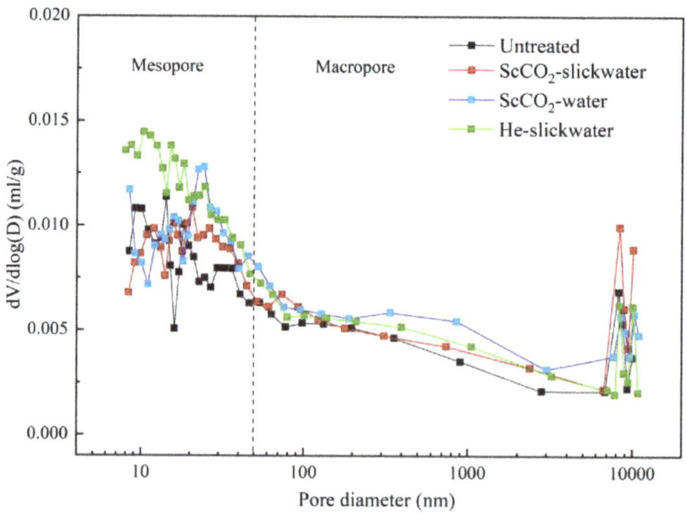

Figure 7. Pore size distributions before and after treatment based on MIP.

Table 4. The SSA, PV, and Da of the shale pores based on MIP.

Sample	SSA (m^2/g)			PV (cm^3/g)			Da (nm)
	Meso-	Macro-	TSSA	Meso-	Macro-	TPV	
Untreated	1.543	0.165	1.709	0.007	0.009	0.015	36.054
$ScCO_2$–slickwater	1.576	0.187	1.763	0.007	0.010	0.017	38.568
$ScCO_2$–distilled water	1.635	0.200	1.835	0.008	0.011	0.019	40.976
He–slickwater	2.328	0.181	2.509	0.010	0.010	0.020	31.253

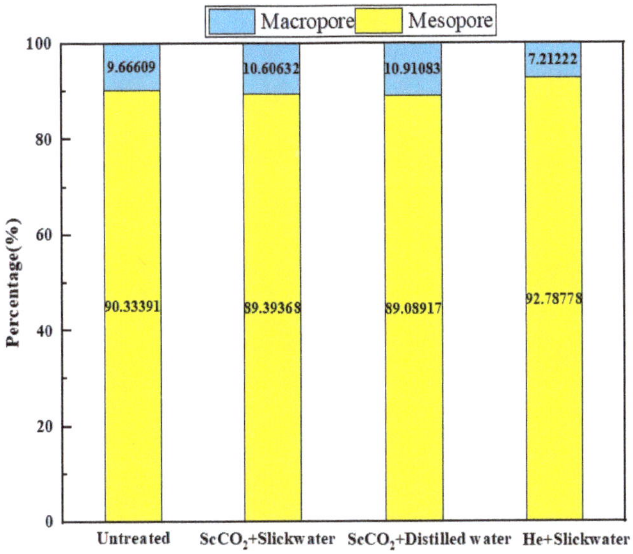

Figure 8. Percentage of specific surface area before and after treatment of mesopores and macropores based on MIP.

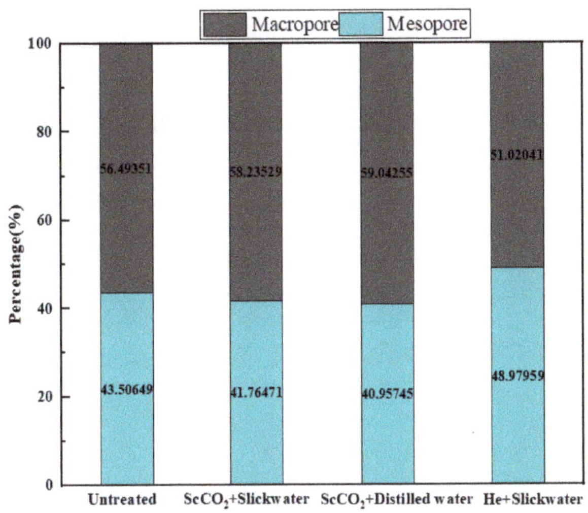

Figure 9. Percentage of PV before and after treatment of mesopores and macropores based on MIP.

3.4. Effect of $ScCO_2$ and Slickwater on the Wettability of Shale Surface

Given that the roughness of the rock surface can significantly influence the accuracy of the contact angle measurement, it is imperative to employ an atomic force microscope to precisely characterize the shale's surface roughness [42]. Figure 10 illustrates the 2D and 3D surface morphology of shale treated by different conditions. It can be found that all the chips' surface roughness was within 1 μm, and the roughness factors of Yanchang shale were 1.0227–1.0453. It is believed that the low surface roughness (<1 μm) is not influential during contact angle measurements. Hence, the roughness did not influence the wettability tests in this study [43].

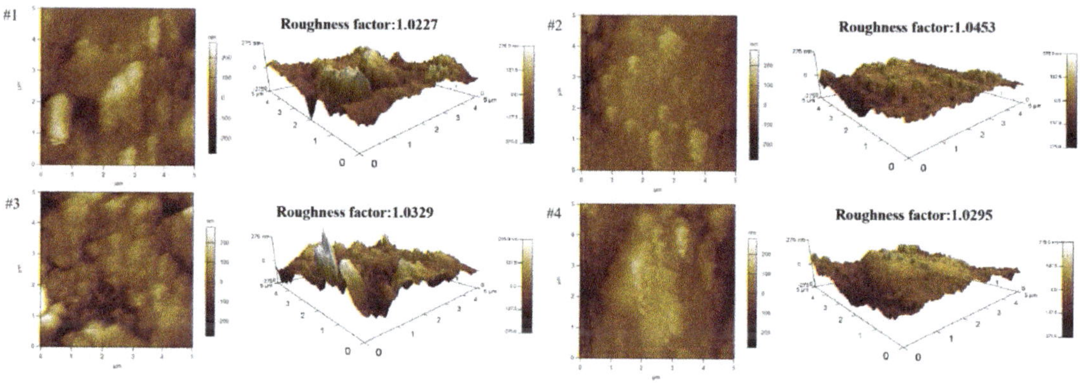

Figure 10. Topography of Yanchang shale measured by AFM. #1: untreated sample, #2: $ScCO_2$–distilled water-treated sample, #3: $ScCO_2$–slickwater-treated sample, #4: He–slickwater-treated sample.

Figure 11 shows the results of contact angle measurements. The contact angles of the untreated shale were 50°–60°, indicating that the raw samples' surface was intermediate water-wet. After the $ScCO_2$–water treatment, the shale's water wettability was significantly vitiated (contact angles: 70°–80°). As rock wettability is closely associated with chemical structures or mineral contents and surface morphology, the AFM results suggest that the influence of morphology on wettability can be trivial, and the mineral contents can be the dominant influence factor [44]. Previous studies and the results of XRD in this work have shown that clay minerals and carbonates can be decreased in the CO_2–water–shale system, and the amount of clay minerals and carbonates determines the hydrophilicity of the rock surface, which explicates the variation of contact angle measurements [45].

Additionally, the contact angles of the samples treated by $ScCO_2$ and slickwater were 60°–70°, indicating that slickwater may deter the water wettability from weakening due to CO_2. Slickwater consists of network and crosslinked-structure PAM, which can be absorbed on the shale's surface. Earlier work found that PAM could interact with CO_2, which may prevent rock exposure to CO_2 and mineralization, which is in accord with the results of XRD in this work [9].

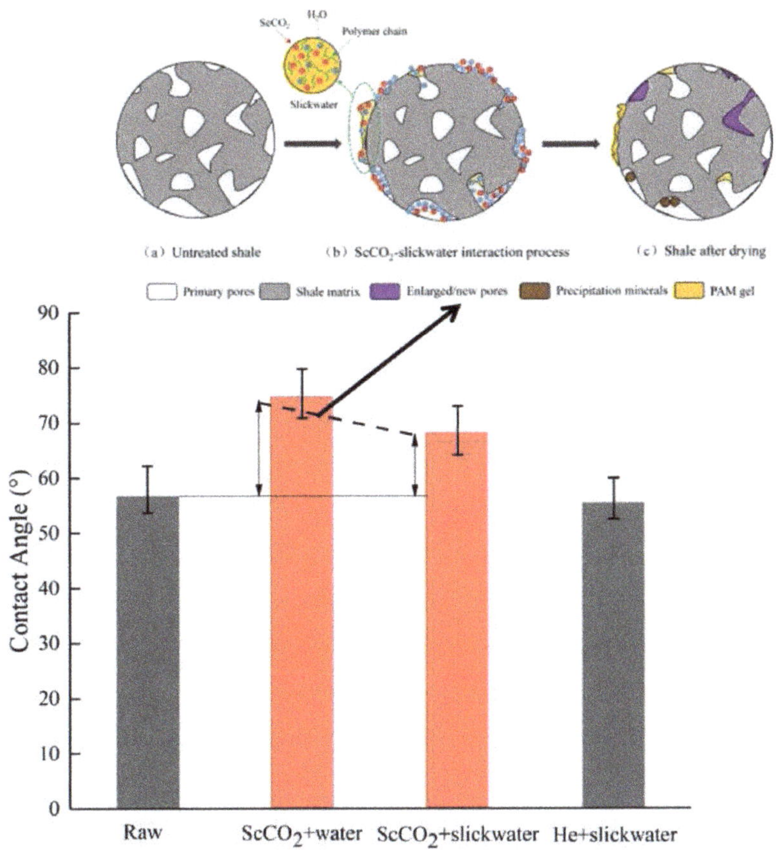

Figure 11. Water contact angles on shale surface are used in different treatment conditions. Absorbed PAM on the shale surface may be the main reason for the lower contact angles of the shale treated by $ScCO_2$ + slickwater than that treated by $ScCO_2$ + water. Modified from [liujie].

3.5. *Implications for the CO_2 Geological Sequestration and Shale Gas Recovery*

Previous studies have clarified that the first decade's security of CO_2 geo-storage can be determined by the capillary force, which is related to the wettability of the rock surface, liquid surface tension, and capillary radius [46]. This can be characterized by the following formula:

$$P = (\rho_{\text{brine}} - \rho_{\text{co}_2})\text{gh} = \frac{2\gamma cos\theta_{\text{co}_2/\text{water}/\text{solid}}}{r}$$

where P is the capillary force (mN/m^2), γ is the surface tension of the liquid (mN/m), and here it is the water surface tension, γwater = 72.8 mN/m, r is the capillary force, and in this study, it can be regarded as the average pore radius of the shale based on LP-NA tests. According to the results of the contact angle measurements and LP-NA, the change of the capillary force of the shale treated by different conditions is shown in Figure 12. Compared with the raw samples, shale samples treated by $ScCO_2$ suffered a massive loss of capillary force. Since the capillary force controls the efficiency of the structural trapping, the CO_2–water–shale systems impair the trapping capability of Yanchang shale. Although using slickwater decreased the negative effect of CO_2 on the shale capillary force, it may still be in an inferior position against the buoyancy of the CO_2 plume in the reservoirs. Therefore, it is necessary to find solutions to change the capillary force of shale artificially in the future. At the same time, the change of the capillary force will not only affect the sealing stability of shale but also influence unconventional natural gas recovery [45]. Reducing the capillary force would decrease the driving force of the water flow in shale pores, which may undermine the blocking effect of water on shale pore structures. Since the blocking effect can be detrimental to the shale gas flow in the shale pores, the change of the capillary force can be conducive to the shale gas recovery [45].

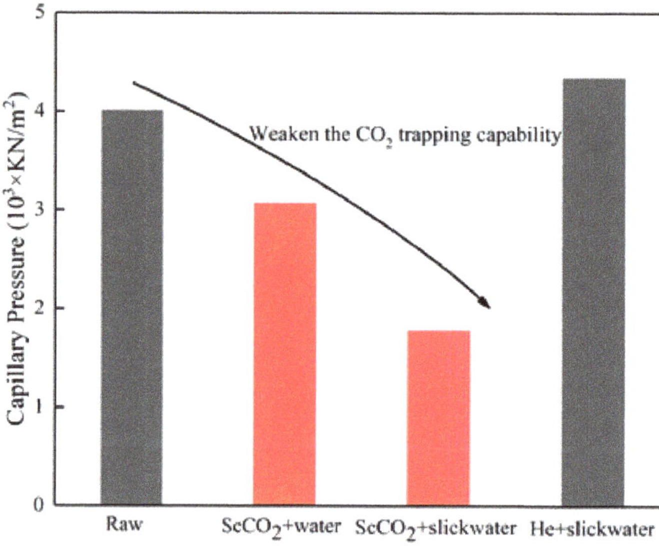

Figure 12. Capillary pressure of shale samples treated by different conditions.

4. Conclusions

In this study, low-pressure N_2 adsorption tests and Mercury intrusion porosimetry tests provide information on the change of the pore structures of Yanchang $ScCO_2$–slickwater-treated shale. In addition, contact angle measurements were used to observe the surface wetting behavior to facilitate the investigation of the capillary force of shale. Our findings indicate that after the $ScCO_2$–slickwater treatment, pore connectivity was impaired due to more micro- and mesopores emerging. The fractal dimensional analysis also suggests that the hybrid treatment strengthened the roughness with pore channels. Based on the XRD results, we assume that the mechanism of pore characteristic changes may be ascribed to the adsorption of PAM and mineralization. This was indirectly proved by the shale surface wetting behavior after the $ScCO_2$–slickwater treatment, as the network and crosslinked-structure PAM can deter the water wettability weakening caused by exposure to $ScCO_2$. Therefore, using slickwater can decrease the negative effect of CO_2 on the shale capillary force to benefit the shale gas recovery, but it may negatively affect the security of CO_2 geo-sequestration.

Author Contributions: C.Z.: Conceptualization, Funding acquisition, Supervision. Q.L.: Conceptualization, Funding acquisition, Supervision. Y.L.: Formal analysis, Investigation, Methodology, Visualization, Writing—original draft. J.T.: Conceptualization, Funding acquisition, Supervision. Y.J.: Conceptualization, Funding acquisition, Supervision. T.G.: Conceptualization, Formal analysis, Investigation, Writing—original draft. All authors have read and agreed to the published version of the manuscript.

Funding: This research received no external funding.

Data Availability Statement: Data will be made available on request.

Conflicts of Interest: The authors declare that they have no known competing financial interests or personal relationships that could have appeared to influence the work reported in this paper.

References

1. Meakin, P.; Huang, H.; Malthe-Sørenssen, A.; Thøgersen, K. Shale gas: Opportunities and challenges. *Environ. Geosci.* **2013**, *20*, 151–164. [CrossRef]
2. Shakya, S.; Li, B.; Etienne, X. Shale revolution, oil and gas prices, and drilling activities in the United States. *Energy Econ.* **2022**, *108*, 105877. [CrossRef]
3. Zhong, C.; Hou, D.; Liu, B.; Zhu, S.; Wei, T.; Gehman, J.; Alessi, D.S.; Qian, P.-Y. Water footprint of shale gas development in China in the carbon neutral era. *J. Environ. Manag.* **2023**, *331*, 117238. [CrossRef]
4. Sakhaee-Pour, A.; Bryant, S.L. Gas permeability of shale. *SPE Reserv. Eval. Eng.* **2012**, *15*, 401–409. [CrossRef]
5. Clarkson, C.; Haghshenas, B.; Ghanizadeh, A.; Qanbari, F.; Williams-Kovacs, J.D.; Riazi, N.; Debuhr, C.; Deglint, H.J. Nanopores to megafractures: Current challenges and methods for shale gas reservoir and hydraulic fracture characterization. *J. Nat. Gas Sci. Eng.* **2016**, *31*, 612–657. [CrossRef]
6. Zhang, X.; Zhu, W.; Xu, Z.; Liu, S.; Wei, C. A review of experimental apparatus for supercritical CO_2 fracturing of shale. *J. Pet. Sci. Eng.* **2022**, *208*, 109515. [CrossRef]
7. Memon, S.; Feng, R.; Ali, M.; Bhatti, M.A.; Giwelli, A.; Keshavarz, A.; Xie, Q.; Sarmadivaleh, M. Supercritical CO_2-Shale interaction induced natural fracture closure: Implications for scCO_2 hydraulic fracturing in shales. *Fuel* **2022**, *313*, 122682. [CrossRef]
8. Xie, W.; Chen, S.; Wang, M.; Yu, Z.; Wang, H. Progress and prospects of supercritical CO_2 application in the exploitation of shale gas reservoirs. *Energy Fuels* **2021**, *35*, 18370–18384. [CrossRef]
9. Lu, Y.; Liu, J.; Tang, J.; Ao, X.; Li, H.; Zhou, J.; Sun, X. Pore changes of slickwater-containing shale under supercritical CO_2 treatment. *Fuel* **2022**, *312*, 122775. [CrossRef]
10. Ao, X.; Wang, B.; Qu, H.; Xiang, Z.; Luo, Z. Swelling of shales with slickwater in carbon dioxide. *Energy Fuels* **2021**, *35*, 5122–5129. [CrossRef]
11. Dayan, A.; Stracener, S.M.; Clark, P.E. Proppant transport in slick-water fracturing of shale-gas formations. In Proceedings of the SPE Annual Technical Conference and Exhibition, New Orleans, LA, USA, 4–7 October 2009.
12. Ribeiro, L.H.; Li, H.; Bryant, J.E. Use of a CO_2-hybrid fracturing design to enhance production from unpropped-fracture networks. *SPE Prod. Oper.* **2017**, *32*, 28–40. [CrossRef]
13. Qin, C.; Jiang, Y.; Zhou, J.; Song, X.; Liu, Z.; Li, D.; Zhou, F.; Xie, Y.; Xie, C. Effect of supercritical CO_2 extraction on CO_2/CH_4 competitive adsorption in Yanchang shale. *Chem. Eng. J.* **2021**, *412*, 128701. [CrossRef]
14. Jiang, K.; Ashworth, P.; Zhang, S.; Hu, G. Print media representations of carbon capture utilization and storage (CCUS) technology in China. *Renew. Sustain. Energy Rev.* **2022**, *155*, 111938. [CrossRef]
15. Isah, A.; Arif, M.; Hassan, A.; Mahmoud, M.; Iglauer, S. A systematic review of Anhydrite-Bearing Reservoirs: EOR Perspective, CO_2-Geo-storage and future research. *Fuel* **2022**, *320*, 123942. [CrossRef]
16. Lu, Y.; Liu, Y.; Tang, J.; Jia, Y.; Tian, R.; Zhou, J.; Chen, X.; Xu, Z.; Cheng, Q. Water wettability alternation of CO_2-water-shale system due to nanoparticles: Implications for CO_2 geo-storage. *Int. J. Greenh. Gas Control* **2023**, *124*, 103836. [CrossRef]
17. Liu, Y.; Tang, J.; Jia, Y.; Lu, Y.; Shang, C.; Tian, R.; Sun, X.; Cheng, Q.; Zhao, G. Investigation on Capillary Force Recovery of Nanoparticle-Coated Shale after Supercritical CO_2 Exposure: Implications for CO_2 Sequestration. *Energy Fuels* **2023**, *37*, 13137–13150. [CrossRef]
18. Wang, Y.; Wang, Q.; Hu, Q.; Cheng, H.; Ilavsky, J.; Kuzmenko, I.; Shang, F.; Liu, L. Pore Structure and Wettability of Bossier Shale, East Texas, United States: Insights from Integrated Porosimetry, Scattering, and Imbibition Approaches. *Energy Fuels* **2023**, *37*, 11042–11054. [CrossRef]
19. Khandoozi, S.; Hazlett, R.; Fustic, M. A critical review of CO_2 mineral trapping in sedimentary reservoirs–from theory to application: Pertinent parameters, acceleration methods and evaluation workflow. *Earth-Sci. Rev.* **2023**, *244*, 104515. [CrossRef]
20. Iglauer, S. CO_2–water–rock wettability: Variability, influencing factors, and implications for CO_2 geostorage. *Acc. Chem. Res.* **2017**, *50*, 1134–1142. [CrossRef]
21. Ao, X.; Lu, Y.; Tang, J.; Chen, Y.; Li, H. Investigation on the physics structure and chemical properties of the shale treated by supercritical CO_2. *J. CO_2 Util.* **2017**, *20*, 274–281. [CrossRef]
22. Zhou, J.; Yang, K.; Zhou, L.; Jiang, Y.; Xian, X.; Zhang, C.; Tian, S.; Fan, M.; Lu, Z. Microstructure and mechanical properties alterations in shale treated via CO_2/CO_2-water exposure. *J. Pet. Sci. Eng.* **2021**, *196*, 108088. [CrossRef]
23. Qin, C.; Jiang, Y.; Zuo, S.; Chen, S.; Xiao, S.; Liu, Z. Investigation of adsorption kinetics of CH_4 and CO_2 on shale exposure to supercritical CO_2. *Energy* **2021**, *236*, 121410. [CrossRef]
24. Liu, B.; Teng, J.; Mastalerz, M.; Schieber, J.; Schimmelmann, A.; Bish, D. Compositional control on shale pore structure characteristics across a maturation gradient: Insights from the Devonian New Albany Shale and Marcellus Shale in the eastern United States. *Energy Fuels* **2021**, *35*, 7913–7929. [CrossRef]
25. Wang, Y.; Zhu, Y.; Zhang, R.; Anovitz, L.M.; Bleuel, M.; Liu, S.; Chen, S. SANS coupled with fluid invasion approaches for characterization of overall nanopore structure and mesopore connectivity of organic-rich marine shales in China. *Int. J. Coal Geol.* **2020**, *217*, 103343. [CrossRef]

26. Lu, Y.; Tian, R.; Tang, J.; Jia, Y.; Lu, Z.; Sun, X. Investigating the mineralogical and chemical effects of CO_2 injection on shale wettability at different reservoir temperatures and pressures. *Energy Fuels* **2021**, *35*, 14838–14851. [CrossRef]
27. Sun, Z.; Zhang, H.; Wei, Z.; Wang, Y.; Wu, B.; Zhuo, S.; Zhao, Z.; Li, J.; Hao, L.; Yang, H. Effects of slick water fracturing fluid on pore structure and adsorption characteristics of shale reservoir rocks. *J. Nat. Gas Sci. Eng.* **2018**, *51*, 27–36. [CrossRef]
28. Balashov, V.N.; Guthrie, G.D.; Lopano, C.L.; Hakala, J.A.; Brantley, S.L. Reaction and diffusion at the reservoir/shale interface during CO_2 storage: Impact of geochemical kinetics. *Appl. Geochem.* **2015**, *61*, 119–131. [CrossRef]
29. Ilgen, A.G.; Aman, M.; Espinoza, D.N.; Rodriguez, M.A.; Griego, J.M.; Dewers, T.A.; Feldman, J.D.; Stewart, T.A.; Choens, R.C.; Wilson, J. Shale-brine-CO_2 interactions and the long-term stability of carbonate-rich shale caprock. *Int. J. Greenh. Gas Control* **2018**, *78*, 244–253. [CrossRef]
30. Jung, H.B.; Um, W.; Cantrell, K.J. Effect of oxygen co-injected with carbon dioxide on Gothic shale caprock–CO_2–brine interaction during geologic carbon sequestration. *Chem. Geol.* **2013**, *354*, 1–14. [CrossRef]
31. Gupta, N.; Verma, A. Supercritical Carbon Dioxide Utilization for Hydraulic Fracturing of Shale Reservoir, and Geo-Storage: A Review. *Energy Fuels* **2023**, *37*, 14604–14621. [CrossRef]
32. Mastalerz, M.; Hampton, L.; Drobniak, A.; Loope, H. Significance of analytical particle size in low-pressure N_2 and CO_2 adsorption of coal and shale. *Int. J. Coal Geol.* **2017**, *178*, 122–131. [CrossRef]
33. Yang, K.; Li, B.; Li, J.; Ren, C. Adsorption characteristics and thermodynamics of CH_4, CO_2, and N_2 on shale at different temperatures. *Energy Fuels* **2022**, *36*, 14079–14093. [CrossRef]
34. Hazra, B.; Wood, D.A.; Vishal, V.; Singh, A.K. Pore characteristics of distinct thermally mature shales: Influence of particle size on low-pressure CO_2 and N_2 adsorption. *Energy Fuels* **2018**, *32*, 8175–8186. [CrossRef]
35. Chen, Y.; Wei, L.; Mastalerz, M.; Schimmelmann, A. The effect of analytical particle size on gas adsorption porosimetry of shale. *Int. J. Coal Geol.* **2015**, *138*, 103–112. [CrossRef]
36. Liang, L.; Xiong, J.; Liu, X. An investigation of the fractal characteristics of the Upper Ordovician Wufeng Formation shale using nitrogen adsorption analysis. *J. Nat. Gas Sci. Eng.* **2015**, *27*, 402–409. [CrossRef]
37. Li, Z.; Shen, X.; Qi, Z.; Hu, R. Study on the pore structure and fractal characteristics of marine and continental shale based on mercury porosimetry, N_2 adsorption and NMR methods. *J. Nat. Gas Sci. Eng.* **2018**, *53*, 12–21. [CrossRef]
38. Yang, R.; He, S.; Yi, J.; Hu, Q. Nano-scale pore structure and fractal dimension of organic-rich Wufeng-Longmaxi shale from Jiaoshiba area, Sichuan Basin: Investigations using FE-SEM, gas adsorption and helium pycnometry. *Mar. Pet. Geol.* **2016**, *70*, 27–45. [CrossRef]
39. Yang, X.; Guo, S. Comparative analysis of shale pore size characterization methods. *Pet. Sci. Technol.* **2020**, *38*, 793–799. [CrossRef]
40. Li, Y.; Wang, S.; Guo, J.; Gou, X.; Jiang, Z.; Pan, B. Reduced adsorption of polyacrylamide-based fracturing fluid on shale rock using urea. *Energy Sci. Eng.* **2018**, *6*, 749–759. [CrossRef]
41. Xu, J.; Georges, M.M.; Chen, J.; Wang, X. Comparison of adsorption ability of PAM, APAM and CPAM on Sodium-Montmorllonite surface: A molecular dynamics study. *Chem. Phys.* **2023**, *574*, 112051. [CrossRef]
42. Zhou, J.; Yang, K.; Tian, S.; Zhou, L.; Xian, X.; Jiang, Y.; Liu, M.; Cai, J. (CO_2-water-shale interaction induced shale microstructural alteration. *Fuel* **2020**, *263*, 116642. [CrossRef]
43. Lu, Y.; Zhou, J.; Li, H.; Tang, J.; Zhou, L.; Ao, X. Gas flow characteristics in shale fractures after supercritical CO_2 soaking. *J. Nat. Gas Sci. Eng.* **2021**, *88*, 103826. [CrossRef]
44. Qin, C.; Jiang, Y.; Luo, Y.; Xian, X.; Liu, H.; Li, Y. Effect of supercritical carbon dioxide treatment time, pressure, and temperature on shale water wettability. *Energy Fuels* **2017**, *31*, 493–503. [CrossRef]
45. Qin, C.; Jiang, Y.; Zhou, J.; Zuo, S.; Chen, S.; Liu, Z.; Yin, H.; Li, Y. Influence of supercritical CO_2 exposure on water wettability of shale: Implications for CO_2 sequestration and shale gas recovery. *Energy* **2022**, *242*, 122551. [CrossRef]
46. Ali, M.; Pan, B.; Yekeen, N.; Al-Anssari, S.; Al-Anazi, A.; Keshavarz, A.; Iglauer, S.; Hoteit, H. Assessment of wettability and rock-fluid interfacial tension of caprock: Implications for hydrogen and carbon dioxide geo-storage. *Int. J. Hydrogen Energy* **2022**, *47*, 14104–14120. [CrossRef]

energies

Article

Research on the Mechanism and Prediction Model of Pressure Drive Recovery in Low-Permeability Oil Reservoirs

Haicheng Liu [1,2,*] and Binshan Ju [1,*]

1 School of Energy Resources, China University of Geosciences (Beijing), Beijing 100083, China
2 Exploration and Development Research Institute of Sinopec Shengli Oilfield Company, Dongying 257015, China
* Correspondence: 13181988003@163.com (H.L.); jubs2936@163.com (B.J.)

Abstract: China boasts significant reserves of low-permeability oil reservoirs, and the economic and efficient development of these reservoirs plays a crucial role in enhancing oil and gas production. However, the "difficult injection and difficult recovery" issue in low-permeability oil reservoirs is a major challenge. To address this, research is conducted on the mechanism of pressure drive based on the mathematical model of oil-water seepage in low-permeability reservoirs and the model of fracture permeability. The study finds that pressure drive technology, by directly delivering the pressure drive agent deep into the low-permeability reservoir, effectively prevents viscosity loss and adhesion retention of the agent in the near-wellbore area. This technology expands the swept volume, improves oil washing efficiency, replenishes formation energy, and facilitates the gathering and production of scattered remaining oil. For reservoirs with higher permeability, pressure drive yields quick results, and high-pressure water injection can be directly adopted for pressure drive to reduce costs. On the other hand, reservoirs with lower permeability have difficulty in water absorption, and the use of surfactant-based pressure drive can effectively reduce the seepage resistance of the reservoir, enhancing its water absorption capacity and improving development outcomes. Based on the mechanism of pressure drive development, further research is conducted on the production characteristics of pressure drive mines. Addressing the variability in pressure drive effects, big data analysis tools such as SHAP analysis and correlation analysis are employed to evaluate the main controlling factors of pressure drive in both new and old areas. Additionally, non-time series and time series pressure drive production forecasting models are established based on pressure drive data.

Keywords: low-permeability oil reservoirs; main controlling factors; model of fracture permeability; physical simulation; mechanism of pressure drive

1. Introduction

Currently, 70% of global oil production originates from mature oilfields, and they will continue to serve as the primary source of oil output in the future [1,2]. Mature oilfields, having undergone prolonged water injection and multiple chemical enhanced oil recovery (EOR) stages, are characterized by high recovery rates and water cuts, posing significant challenges for further development and exploitation in China's petroleum industry. For instance, the Daqing oilfield's low-to-medium permeability reservoirs exhibit comprehensive water cuts ranging from 93% to 96% and recovery rates of 38% to 42%, with substantial remaining reserves [3–6]. However, these reserves are hindered by poor reservoir quality, scattered remaining oil distributions, and injection–production difficulties, rendering further enhancement of recovery rates challenging. Thus, there is an urgent need for efficient reservoir stimulation technologies that can improve reservoir utilization and mobilize scattered remaining oil [7–12].

Xue Changrong [13] analyzed oil recovery and fracturing techniques for low-permeability reservoirs based on practical work experience, offering insights yet lacking empirical validation and in-depth studies on production efficiency. Ren Zeying [14] demonstrated the application of an integrated geological-engineering approach in the low-permeability turbidite sandstone reservoir in the Niuzhuang Sag of the Dongying Depression, significantly enhancing reservoir characterization, drilling efficiency, and oil displacement. While this technology shows promise, it also complicates operations and may slow production rates. Lan Shuang [15] theoretically explored the potential of micro-nano emulsions as enhanced water injection agents for low-permeability reservoirs, highlighting the need for practical implementation and further investigation. The morphological complexity and orientation disorder of pore space evolution during in-situ thermal reconstruction of oil shale, and their effects on permeability are introduced. Pan Duo [16] introduced various EOR technologies, including polymer flooding, composite flooding, microbial flooding, and in-depth profile control, to improve recovery in low-permeability reservoirs. However, these techniques are complex and technically demanding. Liu Jun [17] paid attention to this issue of morphological complexity and azimuthal disorder of evolving pore space in low-maturity oil shale during in-situ thermal upgrading, and impacts on permeability.

Yu Jinzhu [18] reviewed fracturing techniques and trends for low-permeability oilfields, focusing on the state-of-the-art without addressing specific efficiency issues or proposing innovative solutions. Yang Yong [19] conducted numerical simulations to study the impact of various pressure-driven parameters on production enhancement in low-permeability reservoirs. While it identified optimal parameter combinations, the study did not fully address injection difficulties. The evolution of the anisotropic thermophysical properties of low-maturity oil shales under elevated temperatures has been investigated, and a rational prediction of its impact on the restoration of oil development has been conducted [20]. Zhang, S.Q [21] employed a petrophysical and geochemical investigation-based methodology to conduct a multilithology analysis of the reservoir in the Permian Longtan Formation of the southeastern Sichuan Basin, and provided a framework for pressure-driven exploitation. Li, L. [22] studied the distribution and development methods of coal measure gas resources in China, and pressure-driven exploitation was proposed as an important development approach. Cui Chuanzhi [23] analyzed the effects of induced fractures on injection well pressures during pressure-driven water injection, contributing to the understanding of unstable seepage mechanisms but lacking a comprehensive assessment of the impact on oil recovery.

Conventional chemical flooding in low-to-medium permeability reservoirs suffers from low injection efficiency, poor reservoir utilization, and high chemical loss. Traditional hydraulic fracturing establishes high-flow channels but struggles with energy replenishment and sweep efficiency. Researchers have proposed large-scale pressure-driven fracturing, which outperforms conventional chemical flooding and hydraulic fracturing in field tests [24–27]. However, the coupling mechanisms between hydraulic fractures and chemical flooding, as well as the oil displacement mechanisms during chemical agent elastic flowback, remain unclear [28–30]. Furthermore, a systematic approach to studying fluid flow patterns and enhanced recovery outcomes during pressure-driven processes is lacking, hindering large-scale application [31–34].

Previous scholars have conducted extensive research on enhancing the recovery efficiency of low-permeability and tight reservoirs, covering a wide range of aspects. Among these [35,36], they investigate the computational methods for determining well spacing in CO_2 miscible and immiscible flooding, aiming to optimize the efficiency of CO_2 flooding. They employ 3D numerical simulations to study the dynamic propagation of fractures in hydraulic fracturing of low-permeability reservoirs, providing a theoretical basis for fracturing design [37]. The optimal well spacing deployment patterns for low-permeability lens-shaped lithological reservoirs to optimize well pattern design are employed [38]. They introduce the application of an advanced water injection model in low-permeability reservoirs to enhance the effectiveness of waterflooding [39]. The application of volume fracturing technology in tight oil reservoirs is examined and its potential for improving recovery rates is demonstrated [40]. Overall, these studies provide scientific evidence and technical support for the effective development of low-permeability and tight reservoirs; however, there are still several issues that have not been considered.

The effect of pressure drive oil recovery we have studied is more detailed. A is rich in low-permeability reservoir resources, with a total of 428 well groups implemented in low-permeability reservoirs, resulting in an accumulated oil increase of 371,000 tons. This study will demonstrate that pressure drive in low-permeability reservoirs can effectively solve the problem of "difficult injection and difficult recovery". By establishing a mathematical model for oil-water seepage in low-permeability reservoirs and a model of fracture permeability, we investigate the development mechanism of pressure drive. This addresses the challenges of difficult economic production in low-abundance beach-bar sand reservoirs, the issue of interlayer utilization differences in low-permeability reservoirs, and the general problems of poor sweep efficiency and low recovery rates in low-permeability reservoirs. In view of the differences in the effects of pressure drive, big data analysis methods such as SHAP analysis and correlation analysis are used to evaluate the main controlling factors of pressure drive in new and old areas. A non-time series and time series pressure-driven production capacity prediction model based on pressure-driven data-driven has been established. The non-time series pressure-driven production capacity prediction model based on machine learning has high accuracy and has been verified.

2. Materials and Methods

2.1. Mathematical Model of Oil-Water Seepage in Low-Permeability Reservoirs

The pressure analysis model for pressure-driven water injection wells comprises two primary components: the mathematical model of oil-water seepage under pressure-driven water injection and the fracture propagation model during such injection. Prior to establishing the pressure analysis model, the following assumptions are made: ① Two-dimensional planar flow; ② The reservoir only involves oil-water two-phase seepage, with oil and water being immiscible and individually adhering to Darcy's law; ③ Both rock and fluids are slightly compressible; ④ Heterogeneity and anisotropy of the rock are considered; ⑤ The effect of gravity is neglected; ⑥ The fracture propagation is modeled as a composite linear expansion. Utilizing a quarter of the five-spot well pattern as the research unit, a pressure analysis model for pressure-driven water injection is formulated.

Oil Phase Seepage Equation:

$$\frac{\partial}{\partial x}\left[p_o\frac{k_x k_{ro}}{\mu_o}\left(\frac{\partial p_o}{\partial x}-G_o\right)\right]+\frac{\partial}{\partial y}\left[p_o\frac{k_y k_{ro}}{\mu_o}\left(\frac{\partial p_o}{\partial y}-G_o\right)\right]+q_o=\frac{\partial}{\partial t}(\varphi p_o s_o) \quad (1)$$

Water Phase Seepage Equation:

$$\frac{\partial}{\partial x}\left[p_w\frac{k_x k_{rw}}{\mu_w}\left(\frac{\partial p_w}{\partial x}-G_w\right)\right]+\frac{\partial}{\partial y}\left[p_w\frac{k_y k_{rw}}{\mu_w}\left(\frac{\partial p_w}{\partial y}-G_w\right)\right]+q_w=\frac{\partial}{\partial t}(\varphi p_w s_w) \quad (2)$$

Auxiliary Equation:

$$So+Sw=1 \quad (3)$$

In the equations, x and y represent the lengths along the x-direction and y-direction, respectively, in meters (m); t is time, in seconds (s); K_x and K_y are the absolute permeabilities in the x-direction and y-direction, respectively, of the porous medium in the reservoir, in Darcy (D); K_{ro} and K_{rw} are the relative permeabilities of the oil phase and water phase, respectively; ρ_o and ρ_w are the densities of the oil phase and water phase, respectively, in kilograms per cubic meter (kg/m^3); μ_o and μ_w are the viscosities of the oil phase and water phase, respectively, in milli Pascal-seconds (m Pa·s); φ represents the porosity of the porous medium in the reservoir; q_o and q_w are the oil phase and water phase production rates under standard conditions, respectively, in cubic meters per day (m^3/d); p_o and p_w

are the pressures of the oil phase and water phase, respectively, in megapascals (MPa); G is the threshold pressure gradient, in megapascals per meter (MPa/m); S_o and S_w are the saturation of the oil phase and water phase, respectively.

By combining the initial conditions, internal and external boundary conditions with Equations (1)–(3), a two-phase oil-water seepage model can be constructed.

2.2. Model of Fracture Permeability

The exponential model is currently the most widely used mathematical model for stress sensitivity, which describes the relationship between formation pressure and fracture permeability.

$$K = K_0 e^{-\alpha\sigma} \tag{4}$$

$$\sigma = \sigma_o - p \tag{5}$$

In the equation, K represents the fracture permeability, in Darcy (D); K_0 is the initial fracture permeability, also in Darcy (D); α is the dimensionless stress sensitivity coefficient; σ is the effective stress, in megapascals (MPa); σ_0 is the overburden pressure, in megapascals (MPa); and p is the formation pore pressure, also in megapascals (MPa).

Based on previous studies on the relationship between formation pressure and fracture permeability, a fracture permeability model under pressure-driven conditions is established:

When the formation pressure is in the ascending stage:

$$\begin{array}{l} K = K_i, (p \leq p_c) \\ K = K_c e^{-\alpha_1(\sigma_0 - p)} (p > p_c) \end{array} \tag{6}$$

When the formation pressure is in the descending stage:

$$\begin{array}{l} K = K_{\min} e^{-\alpha_2(\sigma_0 - p)}, (p = p_{\min}) \\ K = K_{\min} (p < p_{\min}) \end{array} \tag{7}$$

In the formula, α_1 and α_2 represent the dimensionless stress sensitivity coefficients obtained through fitting.

2.3. Solution of Pressure-Driven Model

The pressure analysis model for pressure-driven water injection wells describes a fluid–solid coupling and oil-water two-phase nonlinear seepage problem, which is challenging to solve analytically or semi-analytically. Therefore, the model is discretized, and the finite difference method is employed to find its solution. In the differential equations governing the seepage of oil and water phases, the right-hand side terms are discretized using the first-order forward difference, while the left-hand side terms are discretized using the second-order central difference. The unknowns are p_o, p_w, s_o, s_w, with the independent variables being (x, y, t). Expanding the differential equations for oil and water phases through discretization, the difference equations for the oil and water phases are obtained as follows:

$$\begin{array}{l} T_{oy_{j-\frac{1}{2}}} p_{oi,j-1}^{n+1} + T_{ox_{j-\frac{1}{2}}} p_{oi-1,j}^{n+1} - \left(T_{ox_{i+\frac{1}{2}}} + T_{ox_{i-\frac{1}{2}}} + T_{oy_{j+\frac{1}{2}}} + T_{oy_{j-\frac{1}{2}}} + \frac{V_{i,j}\beta_{oi,j}}{\Delta t} \right) p_{oi-1,j}^{n+1} + \\ T_{ox_{j+\frac{1}{2}}} p_{oi+1,j}^{n+1} + T_{ox_{j+\frac{1}{2}}} p_{oi,j+1}^{n+1} = V_{pi,j}\rho_{oi,j} \frac{s_{oi,j}^{n+1} - s_{oi,j}^{n}}{\Delta t} - Q_{oi,j} - \frac{V_{i,j}\beta_{oi,j}}{\Delta t} p_{oi,j}^{n} \end{array} \tag{8}$$

$$\begin{array}{l} T_{wy_{j-\frac{1}{2}}} p_{wi,j-1}^{n+1} + T_{wx_{j-\frac{1}{2}}} p_{wi-1,j}^{n+1} - \left(T_{wx_{j+\frac{1}{2}}} + T_{wx_{j-\frac{1}{2}}} + T_{wx_{j+\frac{1}{2}}} + T_{wx_{j-\frac{1}{2}}} + \frac{V_{i,j}\beta_{wi,j}}{\Delta t} \right) p_{wi.j}^{n+1} + \\ T_{wx_{j+\frac{1}{2}}} p_{wi+1,j}^{n+1} + T_{wx_{j+\frac{1}{2}}} p_{wi+1,j}^{n+1} = V_{pi,j}\beta_{wi,j} \frac{s_{wi,j}^{n+1} - s_{wi,j}^{n}}{\Delta t} - Q_{wi,j} - \frac{V_{i,j}\beta_{wi,j}}{\Delta t} p_{wi.j}^{n} \end{array} \tag{9}$$

In this context,

$$
\begin{aligned}
T_{oy_{j\pm\frac{1}{2}}} &= \frac{\lambda_{oy_{j\pm\frac{1}{2}}}\Delta x_i h}{0.5\left(\Delta y_j+\Delta y_{j\pm 1}\right)}\\
T_{ox_{j\pm\frac{1}{2}}} &= \frac{\lambda_{ox_{j\pm\frac{1}{2}}}\Delta y_j h}{0.5(\Delta x_i+\Delta x_{i\pm 1})}, T_{wy_{j\pm\frac{1}{2}}} = \frac{\lambda_{wy_{j\pm\frac{1}{2}}}\Delta x_i h}{0.5\left(\Delta y_j+\Delta y_{j\pm 1}\right)}\\
T_{wy_{j\pm\frac{1}{2}}} &= \frac{\lambda_{wx_{j\pm\frac{1}{2}}}\Delta y_i h}{0.5\left(\Delta x_j+\Delta x_{j\pm 1}\right)}, \lambda_o = \frac{\rho_o k k_{ro}}{\mu_o}, \lambda_w = \frac{\rho_w k k_{rw}}{\mu_w}\\
\beta_0 &= \rho_o \varphi s_o\left(C_p + C_o\right), \beta_w = \rho_w \varphi s_w\left(C_p + C_w\right)\\
V_{i,j} &= \Delta x_i \Delta y_j h, Q_{oi,j} = q_{oi,j}\Delta x_i \Delta y_i h, Q_{wi,j} = q_{wi,j}\Delta x_i \Delta y_j h\\
V_{pi,j} &= \varphi_{i,j}\Delta x_i \Delta y_j h
\end{aligned}
\tag{10}
$$

In Equations (8) to (9), Δx_i represents the unit grid length in the x-direction, with units of meters (m). Δy_j represents the unit grid length in the y-direction, with units of meters (m). Δt represents the unit time, with units of days (d). $p_{oi,j}^{n+1}$ is the oil phase pressure at node (*i,j*) and time step (*n* + 1), with units of MPa. $s_{oi,j}^{n+1}$ is the oil saturation at node (*i,j*) and time step (*n* + 1). $p_{wi,j}^{n+1}$ is the water phase pressure at node (*i,j*) and time step (*n* + 1), with units of MPa. $s_{wi,j}^{n+1}$ is the water saturation at node (*i,j*) and time step (*n* + 1). C_p is the rock compressibility coefficient. C_o is the oil phase compressibility coefficient. C_w is the water phase compressibility coefficient. $V_{i,j}$ is the volume of grid (*i,j*), with units of cubic meters (m^3). $Q_{oi,j}$,$Q_{wi,j}$ are the production or injection rates of grid (*i,j*), with units of cubic meters per day (m^3/d). $V_{pi,j}$ is the pore volume of grid (*i,j*), with units of cubic meters (m^3). $T_{ox_{i\pm\frac{1}{2}}}, T_{ox_{i\pm\frac{1}{2}}}, T_{oy_{j\pm\frac{1}{2}}}, T_{oy_{j\pm\frac{1}{2}}}$ are the relative permeabilities of the water and oil phases, respectively. Using the IMPES (Implicit Pressure, Explicit Saturation) method to implicitly solve for pressure $s_o = 1 - s_w$, we can let $A = \frac{\rho_o}{\rho_w}$ represent porosity and $\left[KW_f = \frac{W(x,t)^3}{12} \times 10^9(8)\right] \times A$, represent the absolute permeability. By introducing these notations and eliminating the saturation terms from the above equations, we focus on solving for pressure implicitly. Additionally, KW_f represents the fracture conductivity, with units of 10^{-3} μm^2·m.

$$
c_{i,j}p_{oi,j-1}^{n+1} + a_{i,j}p_{oi-1,j}^{n+1} + e_{i,j}p_{oi,j}^{n+1} + b_{i,j}p_{oi+1,j}^{n+1} + d_{i,j}p_{oi,j+1}^{n+1} = f_{i,j} \tag{11}
$$

Wherein:

$$
\begin{aligned}
c_{i,j} &= T_{oy_{j-\frac{1}{2}}} + AT_{wy_{j-\frac{1}{2}}}\\
a_{i,j} &= T_{ox_{i-\frac{1}{2}}} + AT_{wx_{i-\frac{1}{2}}}\\
e_{i,j} &= -\left(T_{ox_{i-\frac{1}{2}}} + T_{ox_{i+\frac{1}{2}}} + T_{oy_{j-\frac{1}{2}}} + T_{oy_{j+\frac{1}{2}}} + \frac{V_{i,j}\beta_{oi,j}}{\Delta t}\right) - A\left(T_{Wx_{i-\frac{1}{2}}} + T_{Wx_{i+\frac{1}{2}}} + T_{Wy_{j-\frac{1}{2}}} + T_{Wy_{j+\frac{1}{2}}} + \frac{V_{i,j}\beta_{Wi,j}}{\Delta t}\right)\\
b_{i.j} &= T_{ox_{i+\frac{1}{2}}} + AT_{wx_{i+\frac{1}{2}}}\\
d_{i.j} &= T_{oy_{j+\frac{1}{2}}} + AT_{oy_{j+\frac{1}{2}}}\\
f_{i,j} &= -\left(Q_{oi,j} + \frac{V_{i,j}\beta_{oi,j}}{\Delta t}p_{oi,j}^{n}\right) - A\left(Q_{wi,j} + \frac{V_{i,j}\beta_{wi,j}}{\Delta t}p_{wi,j}^{n}\right)
\end{aligned}
\tag{12}
$$

At each node, by applying Equation (11) to formulate equations, a pentadiagonal system of equations is obtained. By substituting the initial and boundary conditions into this system of equations, the pressure can be solved. Subsequently, the pressure values obtained are substituted into the water phase difference Equation (9) to explicitly solve for the water saturation. Finally, the fracture length, conductivity, pressure, and saturation distributions are updated, and the simulation calculation for the next time step is carried out, thereby achieving the solution of the pressure analysis model for pressure-driven water injection wells.

In the solution flowchart of the pressure-driven water injection well pressure analysis model depicted in Figure 1, the notations N_x, N_y, N_t, and NN represent the number of grid nodes in the x-direction, the number of grid nodes in the y-direction, the number of time steps, and the total number of iterations, respectively. T denotes the total simulation time.

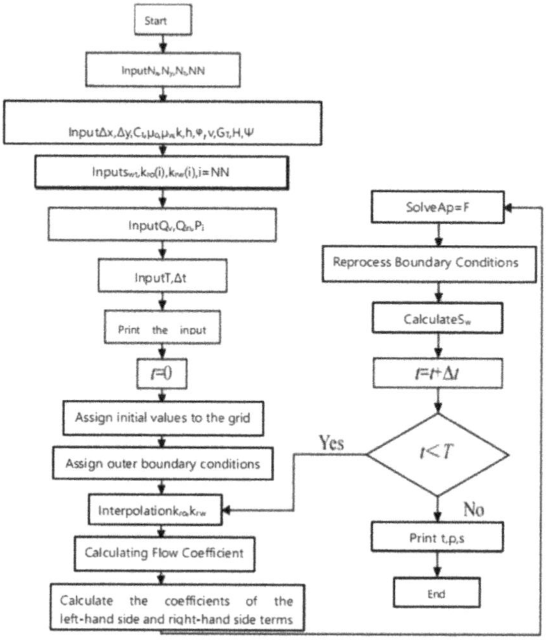

Note: A represents the coefficient matrix term; P represents the pressure term to be solved; F represents the right-hand side pressure matrix term.

Figure 1. Solution Framework for Pressure Analysis Model of Pressure-Driven Water Injection Wells.

During the process of solving the pressure-driven water injection well pressure analysis model, the following steps are typically followed:

(1) Input Basic Parameters: Initially, the fundamental parameters required by the model are input. These parameters may include rock and fluid properties, well specifications, and operational conditions.

(2) Initialization and Boundary Conditions: Each grid cell is assigned an initial value, and the outer boundary conditions of the model are specified. These boundary conditions define the pressure or flow conditions at the edges of the simulated reservoir.

(3) Calculation of k_{ro} and k_{rw}: Using interpolation methods, the relative perm of oil (k_{ro}) and water (k_{rw}) are determined for each grid cell. These values are crucial in determining the flow behavior of the fluids within the reservoir.

(4) Calculation of Flow Coefficients: Based on the reservoir's fundamental parameters and the derived pressure-driven water injection well pressure analysis model, the flow coefficients for each grid cell are computed. These coefficients reflect the ease of fluid flow through the porous medium.

(5) Formation of the Five-Diagonal Equation System: A system of five-diagonal equations is formulated, where each equation represents the pressure balance at a specific

grid cell. This system encapsulates the governing equations of fluid flow and the interactions between adjacent grid cells.

(6) Solution of the Equation System: The five-diagonal equation system is solved numerically to obtain the pressure values at different grid nodes. This step is crucial in determining the pressure distribution within the reservoir under the influence of the water injection well.

(7) Explicit Saturation Calculation and Iteration: Following the pressure calculation, the saturations are updated explicitly using the new pressure values. This step involves applying the appropriate saturation–pressure relationships for the oil and water phases. The updated saturations are then used in the next iteration of the simulation.

(8) Iteration and Simulation Completion: The above steps are repeated in a loop until the end of the simulation time is reached. Each iteration refines the pressure and saturation distributions, providing a more accurate representation of the reservoir's response to the water injection well.

By following this iterative process, the pressure-driven water injection well pressure analysis model enables the prediction of pressure and saturation changes within the reservoir, facilitating informed decision-making in reservoir management and optimization.

3. Mechanism of Enhanced Oil Recovery Through Pressure-Driven Technology

3.1. High Pressure Coefficient: The Key to Establishing a Stable and Continuous Pressure Profile for Long-Term Stable Production

The pressure-driven development of low-permeability reservoirs can create a complex fracture network through high-pressure fluid loss, significantly enhancing permeability and injection capacity for rapid energy replenishment. Compared to conventional water injection, the water injection capacity improves by 10–30 times after pressure-driven water injection. In the pressure-driven modification zone, the pressure front advances, increasing the displacement pressure gradient between wells by 5 to 10 times. Consequently, the range of easily flowing zones expands by 3 to 5 times (Figure 2), enabling the establishment of effective displacement between injection and production wells.

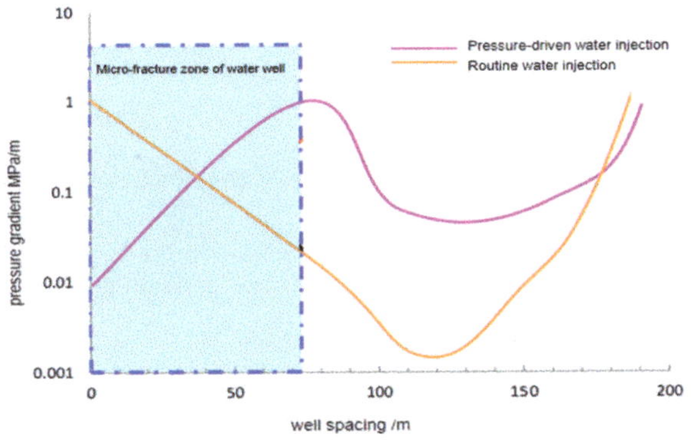

Figure 2. Pressure Gradient Profiles of Conventional Water Injection versus Pressure-Driven Water Injection.

Figure 2 is a comparison of pressure gradient profiles between conventional water injection and pressure-driven water injection. As can be seen from the figure, when the well spacing reaches 80 m, the pressure head moves forward by 800 m, and at this point, the pressure gradient increases by 60 times. This indicates that at this specific well spacing, pressure-driven water injection can significantly improve the pressure gradient, which may enhance the development effect of the reservoir.

There are several reasons for this:

(1) Fluid dynamics effect: As the well spacing increases, the flow path of the fluid becomes longer, leading to an increase in fluid resistance. In this case, using pressure-driven water injection can overcome this resistance, allowing the fluid to pass through the formation more effectively.
(2) Influence of microfracture zones: In the area of microfracture zones between wells, the permeability of the formation will change. Pressure-driven water injection can activate these microfractures by increasing the injection pressure, thereby further improving the fluid flow capacity.
(3) Influence of matrix properties: Factors such as matrix permeability and well spacing may affect the change in pressure gradient. Higher matrix permeability may make it easier for fluids to flow through the formation, while longer well spacing may require higher pressure to maintain the same flow rate.

In the actual oilfield development process, engineers typically determine the best water injection strategy based on field test data and experience. The effect of pressure-driven water injection within this range has been verified through multiple experiments and practical applications.

Our proprietary long-core pressure transmission experiments have demonstrated that the pressure transmission velocity is solely correlated with reservoir physical properties. In contrast, the medium transmission velocity is influenced by both physical properties and pressure gradients. Enhancing physical properties and displacement pressure gradients can reduce the discrepancy between these two velocities.

Pressure transmission refers to the energy transfer resulting from fluid and matrix compression, which is solely related to rock properties and the comprehensive compressibility coefficient of the reservoir. The calculation formula for pressure transmission is:

$$\eta = \frac{k}{\mu} \cdot \frac{1}{\phi \cdot C_t} \tag{13}$$

η: Pressure Transmission Coefficient.
k: Permeability, with units expressed in Darcies (Darcy, darcies) or millidarcies (mD).
μ: Dynamic Viscosity, quantified in units of Pascal-seconds (Pa·s) or centipoise (cP), where 1 Pa·s is equivalent to 1000 cP.
φ: Porosity, which represents the ratio of pore space to the total volume, is a dimensionless value ranging from 0 to 1.
Ct: Total compressibility factor, encompassing the compressibility of both the fluid and the rock matrix, measured in units of reciprocal Pascals (1/Pa).

Medium transmission refers to the mass transfer caused by fluid migration, which is solely related to rock properties, fluid properties, and displacement pressure gradients. The calculation formula for medium transmission is:

$$\mathrm{v} = \frac{k}{\mu} \cdot \frac{\Delta p}{L} \tag{14}$$

K: Permeability, units: Darcy (Darcy, darcies) or millidarcies (millidarcies, mD).
μ: Dynamic Viscosity, units: Pascal-seconds (Pa·s) or centipoise (centipoise, cP), where 1 Pa·s = 1000 cP.
Δp: Pressure Gradient, units: Pascals (Pa) or bars (bar), where 1 bar $\approx$ 100,000 Pa.
v: Fluid Flow Velocity, units: meters per second (m/s) or other velocity units.
L: Medium length, meters (m).

Physical simulation experiments on pressure transmission and medium transmission indicate that the lower the reservoir permeability, the greater the discrepancy between the pressure transmission velocity and the medium transmission velocity (Figure 3). When the permeability increases from 5×10^{-3} μm^2 to 35×10^{-3} μm^2, the pressure transmission velocity increases from 70 cm/min to 80 cm/min, a mere increase of 10 cm/min. However, under a pressure gradient of 0.025 MPa/m, the medium transmission velocity increases from approximately 20 cm/min to 50 cm/min. At a pressure gradient of 0.015 MPa/m, the medium transmission velocity rises from around 5 cm/min to 35 cm/min, and under a pressure gradient of 0.01 MPa/m, it increases from approximately 2 cm/min to 25 cm/min. Furthermore, the lower the reservoir permeability, the more significant the discrepancy between the pressure transmission velocity and the medium transmission velocity becomes.

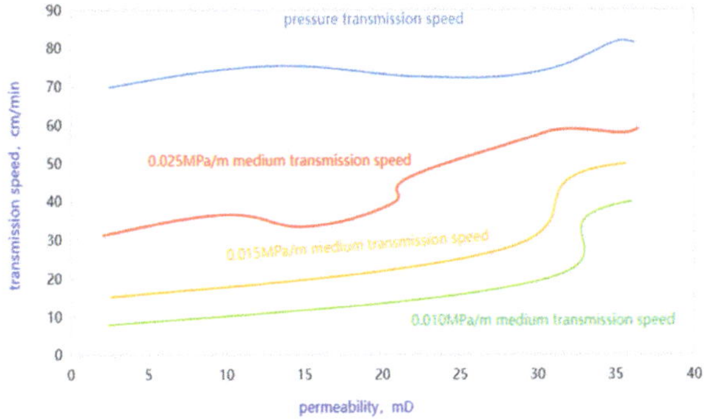

Figure 3. Variation Curves of Pressure and Fluid Transmission Velocity with Permeability.

Field practice in pressure-driven oil recovery demonstrates that under conditions of high pressure coefficients, the oil pressure drop rates of both oil and water wells are essentially the same, maintaining a relatively constant injection–production pressure differential. This is conducive to establishing a continuous and stable pressure profile, resulting in a relatively high initial production, a long natural flow period, and minimal decline.

Continuous Pressure Transmission Type: After the first phase of pressure-driven water injection in D-1, the pressure coefficient reached 1.8, with an initial daily oil production of 10.4 t/d. The natural flow period has been prolonged (>220 days) and is still ongoing without any decline. The oil pressure drop rate of the water well is 0.07 MPa/d, while that of the oil well is 0.1 MPa/d (Figure 4a, Continuous Pressure Transmission Type).

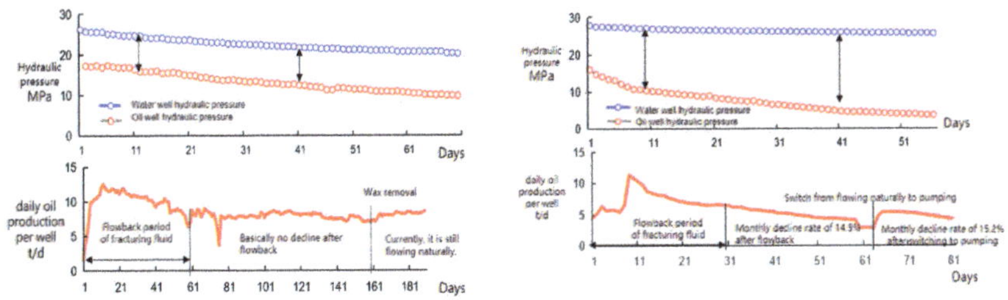

(a) (b)

Figure 4. Variation Curves of Oil Pressure and Daily Oil Production of Oil and Water Wells After Well Opening in Block D-1-1. (**a**) Continuous Pressure Transmission Type, (**b**) Unstable Pressure Transmission Type.

Unstable Pressure Transmission Type: After pressure-driven water injection in Block E-1, the pressure coefficient reached 1.3, with an initial daily oil production of 7.3 t/d. However, the natural flow period was short (62 days), followed by a monthly decline rate of 14.5% after flowback and 15.2% after switching to pumping, indicating a significant decline. The oil pressure drop rate of the water well was 0.04 MPa/d, while that of the oil well was 0.2 MPa/d (Figure 4b. Unstable Pressure Transmission Type).

Numerical simulation studies on fluid–solid coupling in pressure-driven oil reservoirs have shown that a high pressure coefficient can ensure stable pressure fronts, enhance the displacement pressure gradient, and facilitate the establishment of a stable and continuous pressure profile (Figure 5). The numerical simulation of the pressure-driven reservoir indicates that a high pressure coefficient ensures a stable pressure head, increases the displacement pressure gradient, and is beneficial for the establishment of a stable and continuous pressure profile. The advanced pressure drive forms a continuous pressure profile between the oil and water wells. Moreover, the high-pressure system can achieve high and stable production with long-term natural flow. Therefore, three regions will form: (1) In the natural flow zone (blue line), the pressure transmission profile is stable, with continuous natural flow; (2) In the stable pumping zone (yellow line), the displacement pressure is stably established, and the pumping maintains certain production capacity; (3) In the depletion decline zone (green line), near the original reservoir pressure, the production declines rapidly, with no period of natural flow.

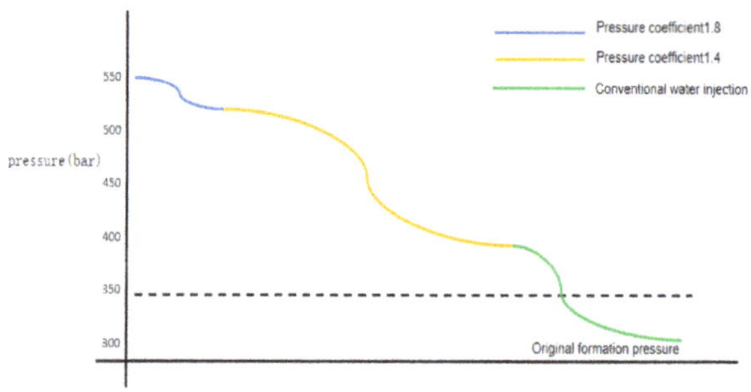

Figure 5. Continuous Pressure Profiles between Oil and Water Wells under Different Pressure Coefficients for Advanced Pressure-Driven Injection.

Based on the above research, it is concluded that a high-pressure system can achieve long-term high and stable production with continuous natural flow.

3.2. Sequential Pressure Drive Followed by Fracturing Facilitates the Formation of Complex Redirected Fracture Networks

Physical simulations indicate that high pore pressure reduces the compressive strength of rocks, making reservoirs more susceptible to fracturing. Higher reservoir pressure coefficients lead to lower rock fracture pressures, favoring the development of complex fracture networks. When water injection wells undergo pressure drive prior to fracturing oil wells, the stress field changes, resulting in the formation of redirected fractures.

A high pressure coefficient can ensure a stable pressure front, enhance the displacement pressure gradient, and help establish a continuous pressure profile. When a water injection well is pressure driven before fracturing an oil well, it can cause changes in the stress field. This change may cause the pressure front to become unstable. So, after regular water injection, sudden pressure changes can lead to an unstable pressure head, which in turn affects pressure distribution and causes a decrease. In addition, high pore pressure reduces the compressive strength of rocks, making reservoirs more prone to fracture and forming complex fracture networks. This further affects the propagation and distribution of pressure, leading to instability of the pressure front.

Rock mechanics failure experiments on cores from Block E-1-2 under varying pore pressures reveal that increased pore pressure decreases compressive strength. When the pore pressure is 10 MPa, the rock fracture strain variation rate is approximately 1.0, with an axial stress difference of around 100 MPa. At a pore pressure of 30 MPa, the rock fracture strain variation rate decreases to approximately 0.5, with an axial stress difference of approximately 25 MPa (Figure 6).

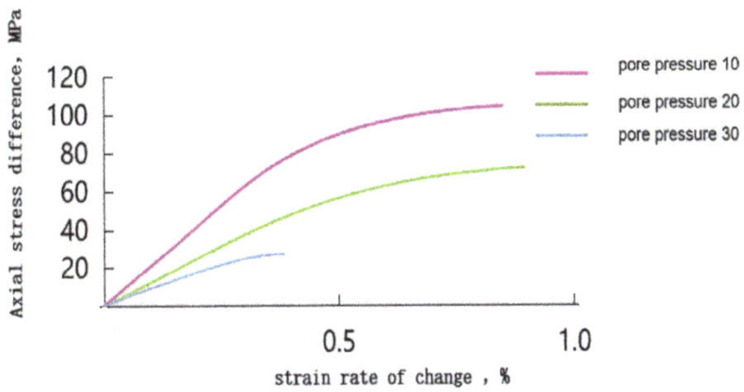

Figure 6. Rock Mechanics Failure Curves of Cores from Block E-1-2 under Different Pore Pressures.

Fracture strength experiments on cores from Block E-1-2 under varying pressure coefficients demonstrate that higher pressure coefficients result in lower fracture pressures. At a pressure coefficient of 1.2, the core fracture pressure is 104.8 MPa; at 1.5, it decreases to 92.9 MPa; and at 1.8, it drops further to 30 MPa (Figure 7).

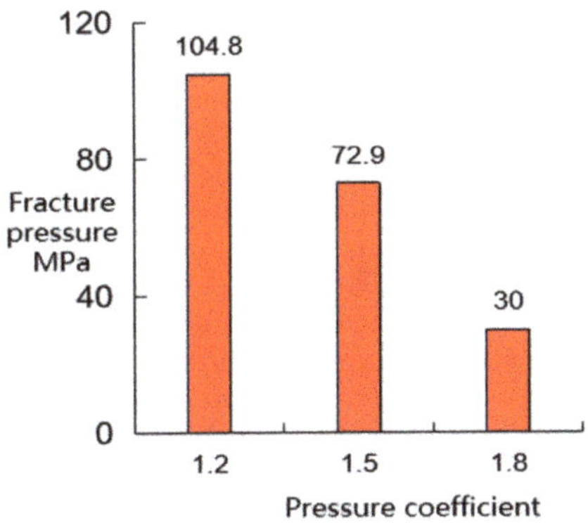

Figure 7. Histogram of Fracture Pressures of Cores from Block E-1-2 under Different Pressure Coefficients.

In the instance where the lithostatic ratio (LSR) is 1.2, the reduction in the fracturing pressure of the rock to 30 MPa is attributable to several factors:

(1) Influence of Pore Pressure: The increase in pore pressure leads to a decrement in the compressive strength of the rock, rendering it more susceptible to fracturing. This is corroborated by Figure 6, wherein the stress differential curves under varying pore pressures illustrate a progressive decline in the rock's load-bearing capacity as pore pressure intensifies.

(2) Variations in Formation Stress: The alteration of the lithostatic ratio impacts the effective stress state of the formation. When the formation is under-compacted (with a lithostatic ratio less than 1), the rock may exhibit higher compressive strength; conversely, as the lithostatic ratio approaches or exceeds unity, the increment in pore fluid pressure results in a reduction in the rock's effective stress, thereby diminishing its resistance to fracturing.

(3) Support from Physical Simulation Experimentation: The relationship between the lithostatic ratio and fracturing pressure depicted in Figure 8 further endorses this perspective. As the lithostatic ratio increases from 1.0 to 1.8, a marked decrease in the rock's fracturing pressure is observed, with a significant drop to 72.9 MPa at a lithostatic ratio of 1.5, and a further decline to 30 MPa when the ratio is 1.2.

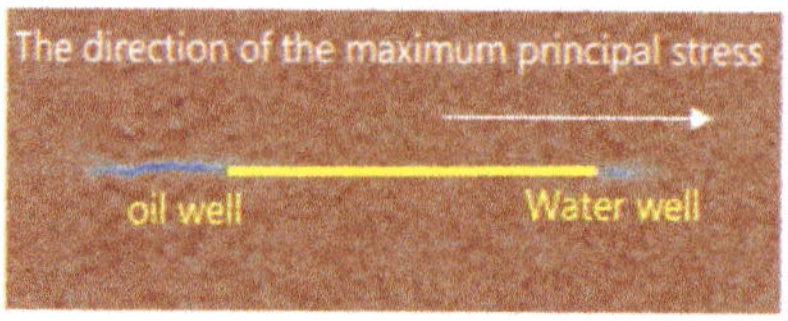

(a)

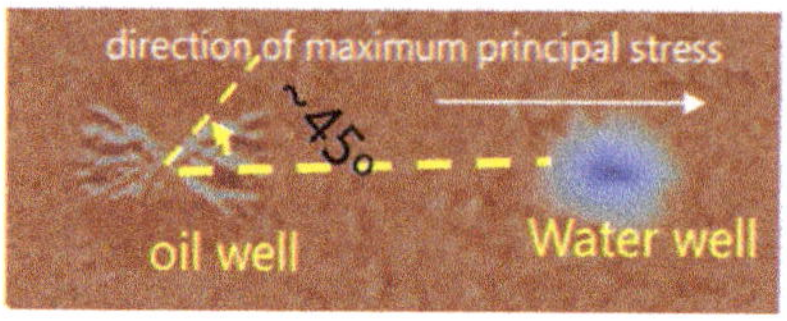

(b)

Figure 8. Schematic Diagram of Fracture Propagation in Oil Wells after Initially Fracturing Water Wells and Subsequently Applying Pressure-Driven Injection to Oil Wells, and Schematic Diagram of Fracture Propagation in Oil Wells after Initially Applying Pressure-Driven Injection to Water Wells and Subsequently Fracturing Oil Wells. (**a**) Oil wells are fractured first, followed by pressure drive for water wells. (**b**) Water wells undergo pressure drive first, followed by fracturing for oil wells.

The decrease in fracturing pressure to 30 MPa at a lithostatic ratio of 1.2 is attributed to the augmentation of pore pressure, which precipitates a decline in the rock's compressive strength, concurrently with the pivotal role played by changes in formation stress. These findings are substantiated by the results of physical simulation experiments and are in alignment with actual field production scenarios in oil and gas operations.

Numerical simulations of firstly fracturing an oil well followed by pressure drive of a water well indicate that when an oil well is fractured and put into production while maintaining a formation pressure coefficient of 1.0, subsequent pressure drive of the water well tends to induce the formation of long fractures in the oil well that are parallel to the principal stress direction, creating wings on both sides (Figure 8a. Oil wells are fractured first, followed by pressure drive for water wells).

Physical experiments involving water well pressure drive preceding oil well fracturing demonstrate that when the water well undergoes water injection and pressure drive two months in advance, resulting in a formation pressure coefficient reaching 1.8, subsequent fracturing of the oil well leads to the formation of complex fractures, with fracture directions deflecting by approximately 45° (Figure 8b. Water wells undergo pressure drive first, followed by fracturing for oil wells).

Field practice has shown that when fracturing precedes pressure drive, rapid water breakthrough occurs in some oil wells. Adjusting the sequence of production and injection, i.e., performing pressure drive before fracturing, can effectively control water breakthrough.

In Block A-1, the water injection well was fractured prior to the pressure-driven enhancement of the oil well, resulting in rapid water breakthrough in the oil well. In December 2018, well A-1-1 in this block was fractured with two stages, each having a half-length of 122 m. In January 2021, generalized pressure-driven water injection was performed in the water injection well of this block, with a total cumulative water injection volume of 3.7×10^4 m^3. Following the pressure-driven water injection, the water cut in well A-1-1, aligned with the principal stress direction, increased rapidly, leading to swift water breakthrough in the oil wells in that same direction (Figure 9).

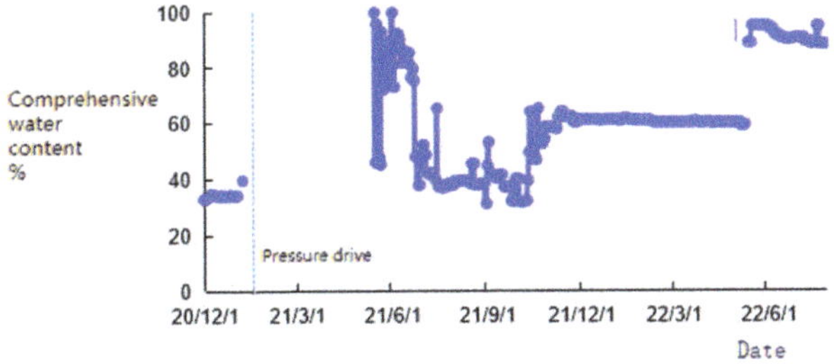

Figure 9. Water Cut Variation Curve of Oil Well A-1-1 in the Direction of Principal Stress (Block A).

In Block Bin 435, water wells were initially subjected to pressure-driven injection prior to fracturing the oil wells, leading to a low water cut and a gradual increase in water cut for the oil wells. In January 2023, the water wells in this block underwent staged pressure-driven injection, with a total cumulative water injection volume of 9.6×10^4 m^3. By April 2023, the oil wells were brought into production following staged fracturing, which included three fracture stages and a half-fracture length of 157 m. After production commenced, the water cut remained stable, with a slow increase observed (Figure 10).

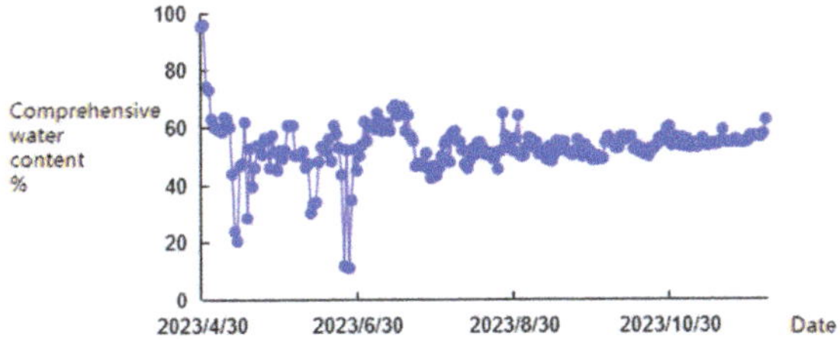

Figure 10. Water Cut Variation Curve of Oil Well B-1-1 in the Direction of Principal Stress (Block B).

In Block B-1, the water injection well underwent pressure-driven treatment prior to the fracturing of the oil well, resulting in a low water cut and a gradual increase in the oil well's water cut. In January 2023, the water injection well in this block experienced staged pressure-driven water injection, with a total cumulative volume of 9.6×10^4 m^3. In April 2023, the oil well was brought into production following staged fracturing, which included three fracture stages and a half-fracture length of 157 m. After production commenced, the water cut remained stable and increased slowly (Figure 10).

3.3. Reservoir Physical Properties and Injection Displacement Are the Main Controlling Factors Affecting the Morphology and Extension Range of Fracture Network

Physical experiments indicate that the rock-breaking mechanisms vary among reservoirs with different physical properties. Generally, fractures induced by pressure drive in low-permeability reservoirs form during the fracture initiation stage, whereas those in ultra-low-permeability reservoirs occur during the fracture propagation stage after rupture. The distinct stages lead to differences in the morphology and scale of pressure-driven fractures in reservoirs with varying physical properties.

Specifically, when the brittleness index is 2.71 and the gas-measured permeability is 16.0×10^{-3} μm^2, pressure-driven fractures are generated during the fracture initiation stage (Figure 11a. Brittleness Index: 2.71 Gas Logging Permeability: 16.0 mD).

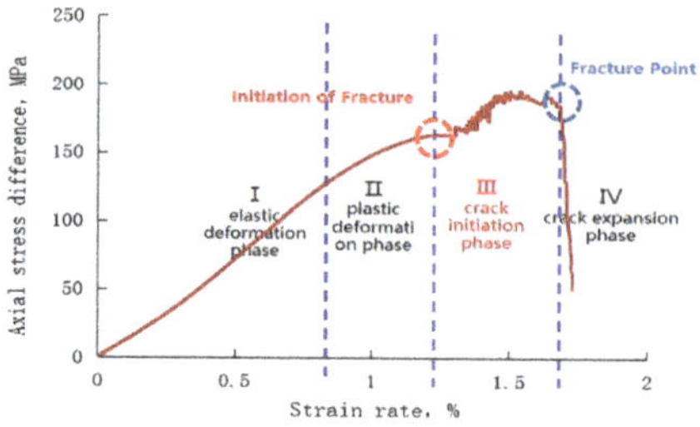

(a)

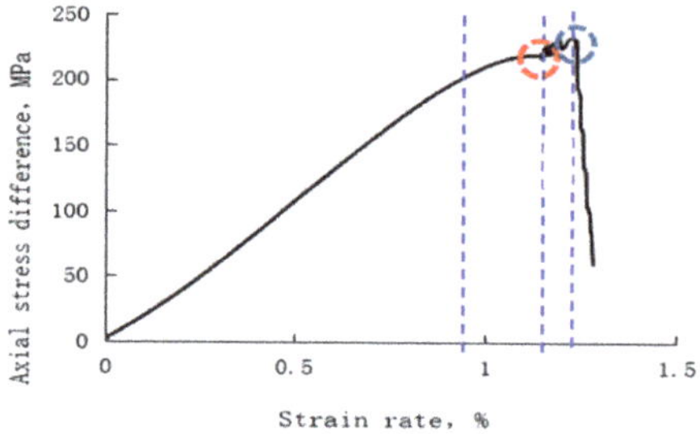

(b)

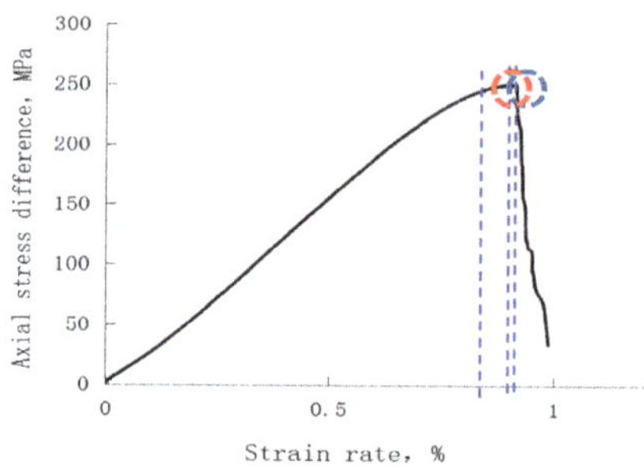

(c)

Figure 11. Strain–stress difference curves at permeabilities of 16.0×10^{-3} um^2, 3.58×10^{-3} um^2, and 0.871×10^{-3} um^2, respectively. (**a**) Brittleness Index: 2.71 Gas Logging Permeability: 16.0 mD. (**b**) Brittleness Index: 59.7 Gas Logging Permeability: 3.58 mD. (**c**) Brittleness Index: 88.4 Gas Logging Permeability: 0.871 mD.

When the brittleness index is 59.7 and the gas-measured permeability is 3.58×10^{-3} μm^2, pressure-driven fractures emerge during the fracture propagation stage following rupture (Figure 11b. Brittleness Index: 59.7 Gas Logging Permeability: 3.58 mD).

With a brittleness index of 88.4 and a gas-measured permeability of 0.871×10^{-3} μm^2, pressure-driven fractures are generated during the fracture propagation stage subsequent to rupture (Figure 11c. Brittleness Index: 88.4 Gas Logging Permeability: 0.87 mD).

Numerical simulations indicate that higher permeability leads to shorter fracture propagation distances and more uniform fracture patterns, while higher injection rates result in longer fracture propagation distances and less uniform fracture patterns. Therefore, to achieve precise control over fracture morphology and propagation distance, it is necessary to match different injection rates with varying reservoir properties.

The schematic diagram of pressure-driven fracture propagation under different perms, with an injection rate of 2×10^{-4} m^3/s and a cumulative injection volume of 0.1 PV, is presented as follows:

When the permeability is 3×10^{-3} μm^2, the fracture propagation distance is the longest, with fractures extending in all directions. In contrast, when the permeability is 50×10^{-3} μm^2, the fracture propagation distance is the shortest, resulting in a more uniform fracture pattern (Figure 12a. Fracture network formed by pressure drive in reservoirs with different physical properties).

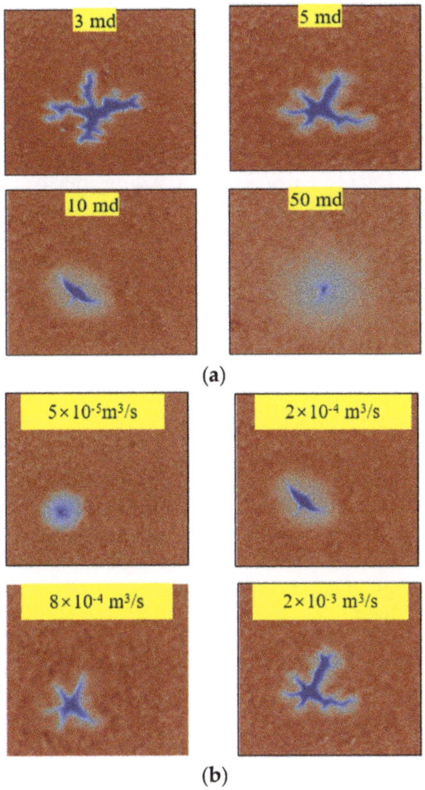

Figure 12. Fracture propagation patterns under different permeabilities, and fracture propagation patterns under different injection rates. (**a**) Fracture network formed by pressure drive in reservoirs with different physical properties. (**b**) Fracture network formed by pressure drive in reservoirs with different physical properties.

With a permeability of 10×10^{-3} μm^2 and a cumulative injection volume of 0.1 PV, the schematic diagram of pressure-driven fracture propagation under different injection rates is shown below:

When the injection rate is 2×10^{-3} m^3/s, the fracture propagation distance is the longest, with fractures extending in all directions. Conversely, at an injection rate of 5×10^{-5} m^3/s, the fracture propagation distance is the shortest, leading to a more uniform fracture pattern (Figure 12b. Fracture network formed by pressure drive in reservoirs with different physical properties).

By implementing advanced high-pressure drive with large well spacing in water wells, we reduced pressure sensitivity and friction, thereby addressing the challenge of inefficient production in low-abundance beach-bar sand reservoirs. Based on the different physical properties of oil layers, we innovatively established a three-dimensional pressure drive approach, optimizing different injection rates to control the propagation morphology of pressure-driven fractures. This achieved high-pressure fluid loss and fracture network creation, resolving the issue of interlayer production differences in ultra-low-permeability reservoirs. Through differential energy replenishment, response induction, displacement adjustment, and flooding, we achieved a synergistic and uniform displacement in the energy field, overcoming the problems of poor sweep efficiency and low recovery rates in conventional low-permeability reservoirs. The in-depth study of pressure drive mechanisms provides guidance for expanding the technical model of pressure drive development in low-permeability reservoirs.

4. Research on Production Characteristics of Pressure Drive in Field Operations

Subsequently, we conducted research on the production characteristics of pressure drive in field operations. Following pressure drive implementation, the daily injection rate per well increased by more than two times, effectively solving the problem of "injection difficulty". The formation pressure maintenance level rose from 69.3% to 95.7%, effectively restoring formation energy. The fluid production capacity of different types of reservoirs improved by more than 1.2 times, addressing the challenge of "production difficulty".

We divided the regions into old and new areas, with the previously studied region designated as the old area and the newly developed one as the new area. Pressure drive effectively overcame the "injection difficulty" issue. In the new area, after pressure drive, the water injection capacity of turbidite rocks increased from 24.6 m^3/d to 69.8 m^3/d, and that of beach-bar sands increased from 42.9 m^3/d to 67.4 m^3/d. The daily water injection capacity after oil well conversion reached 37.9 m^3/d, with a more than two-fold increase compared to pre-pressure drive levels, and a greater improvement observed in turbidite rocks than in beach-bar sands (Figure 13).

In the old area, after pressure drive, the water injection capacity of turbidite rocks increased from 11.8 m^3/d to 29.9 m^3/d, and that of beach-bar sands increased from 7.7 m^3/d to 21.3 m^3/d. The daily water injection capacity after oil well conversion reached 20 m^3/d, also with a more than two-fold increase compared to pre-pressure drive levels, and a greater improvement in turbidite rocks than in beach-bar sands (Figure 13a. Effect diagram of single-well daily injection increased by more than 2 times compared to before pressure driving).

Pressure drive effectively restored formation energy, with notable improvements in pressure maintenance levels across different rock types. Specifically, for turbidite rocks, the pressure maintenance level increased from 69.1% prior to pressure drive to 100.3% post-pressure drive. Similarly, in beach-bar sands, the pressure maintenance level rose from 67.8% before pressure drive to 93.2% after the process. For conglomerates, the pressure maintenance level improved from 73.5% pre-pressure drive to 88.2% post-pressure drive. On average, the pressure maintenance level across all rock types increased from 69.3% before pressure drive to 95.7% afterward, with turbidite rocks exhibiting the highest pressure recovery level, followed by beach-bar sands, and then conglomerates (Figure 13b. Effect diagram showing a pressure maintenance level of 69.3% before pressure driving and recovery to 95.7% after pressure driving).

Pressure drive addressed the challenge of "inadequate production". In the new area, the daily fluid production of turbidite rocks increased from 6.9 t/d before pressure drive to a peak of 8.8 t/d afterward, currently stabilizing at 7.6 t/d. For beach-bar sands, daily fluid production rose from 8.8 t/d pre-pressure drive to a peak of 17.0 t/d post-pressure drive, currently at 8.5 t/d. In conglomerates, daily fluid production increased from 10.0 t/d before pressure drive to a peak of 17.1 t/d afterward, currently averaging 13.8 t/d. These data indicate that turbidite rocks and conglomerates maintained relatively stable fluid production post-pressure drive, while beach-bar sands experienced a sharper increase and subsequent decline (Figure 13c. Effect diagram showing liquid production capacity increased by more than 1.2 times for different types of reservoirs).

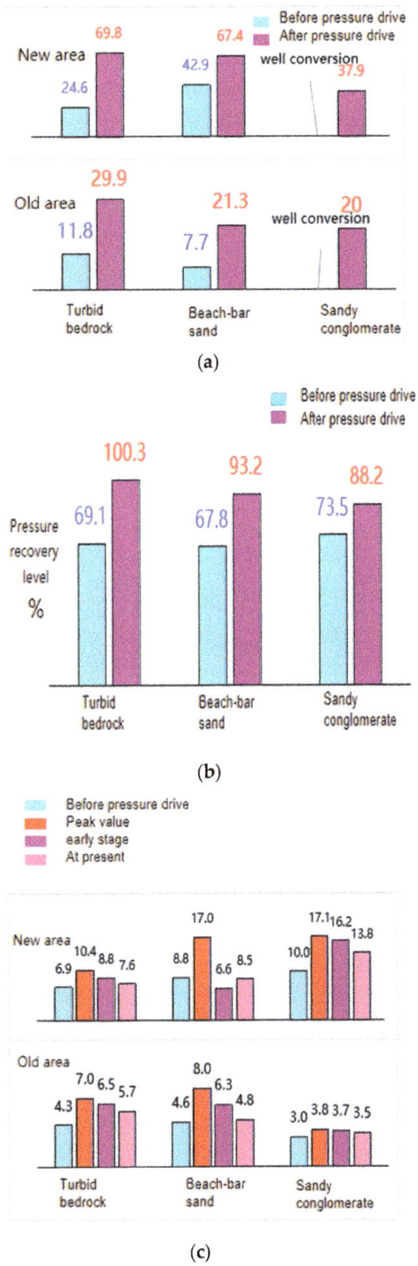

Figure 13. Histogram of Water Injection Capacity after Transition from Pressure-Driven Injection to Conventional Water Injection in Reservoirs of Different Lithologies. (**a**) Effect diagram of single-well daily injection increased by more than 2 times compared to before pressure driving. (**b**) Effect diagram showing a pressure maintenance level of 69.3% before pressure driving and recovery to 95.7% after pressure driving. (**c**) Effect diagram showing liquid production capacity increased by more than 1.2 times for different types of reservoirs.

In the old area, the daily fluid production of turbidite rocks rose from 4.3 t/d before pressure drive to a peak of 7.0 t/d afterward, currently at 5.7 t/d. For beach-bar sands, daily fluid production increased from 4.6 t/d pre-pressure drive to a peak of 8.0 t/d post-pressure drive, currently stabilizing at 4.8 t/d. In conglomerates, daily fluid production rose from 3.0 t/d before pressure drive to a peak of 3.8 t/d afterward, currently averaging 3.5 t/d. These observations suggest that turbidite rocks and conglomerates in the old area also maintained good fluid production capacity post-pressure drive, while beach-bar sands once again demonstrated a steeper increase followed by a quicker decline (Figure 13c. Effect diagram showing liquid production capacity increased by more than 1.2 times for different types of reservoirs).

5. Identify the Main Controlling Factors of Pressure-Driven Technology

In view of the differences in the effects of pressure drive, retrospective evaluations of the main controlling factors of pressure drive in new and old areas are conducted using big data analysis methods such as SHAP analysis and correlation analysis.

The SHAP values for the influencing factors on the effectiveness of pressure drive in the new area are as follows: water injection volume during pressure drive has a SHAP value of 1.1, injection–production well spacing has a SHAP value of 1.08, injection rate has a SHAP value of 0.9, original formation pressure has a SHAP value of 0.66, permeability has a SHAP value of 0.55, reservoir thickness has a SHAP value of 0.3, mid-depth of the reservoir has a SHAP value of 0.22, porosity has a SHAP value of 0.2, and crude oil viscosity has a SHAP value of 0.04 (Figure 14a. SHAP Values for Factors Influencing the Effectiveness of the New Area).

Factors Influencing Cumulative Oil Increment in the New Area and Their Correlation Coefficients:

The correlation coefficients for factors influencing cumulative oil increment in the new area are: water injection volume during pressure drive (0.45), injection–production well spacing (0.43), injection rate (0.40), original formation pressure (0.18), permeability (0.31), reservoir thickness (0.15), mid-depth of the reservoir (0.02), porosity (0.01), and crude oil viscosity (0.02) (Figure 14b Correlation Coefficients of Factors Influencing Cumulative Oil Increment in the New Area).

According to the SHAP values of the influencing factors of the effectiveness status in the new area and the correlation coefficient values of the influencing factors of cumulative oil increase in the new area, the impact values of the main controlling factors of pressure-driven technology in the new area, from largest to smallest, are: pressure-driven water injection volume > injection–production well spacing > injection rate.

SHAP Values for Factors Influencing Effectiveness in the Old Area:

The SHAP values for factors influencing the effectiveness of pressure drive in the old area are: depletion level (1.1), water injection volume during pressure drive (0.9), injection–production well spacing (0.72), injection rate (0.65), permeability (0.36), depleted volume (0.3), water cut before pressure drive (0.28), cumulative injection before pressure drive (0.25), original formation pressure (0.25), mid-depth of the reservoir (0.24), reservoir thickness (0.18), porosity (0.18), and crude oil viscosity (0.11) (Figure 14c. SHAP Values for Factors Influencing the Effectiveness of the Mature Area).

Correlation Coefficients for Cumulative Oil Increment in the Old Area:

The correlation coefficients for factors influencing cumulative oil increment in the old area are: depletion level (0.61), water injection volume during pressure drive (0.53), injection–production well spacing (0.50), injection rate (0.33), permeability (0.33), reservoir thickness (0.28), depleted volume (0.25), water cut before pressure drive (0.13), cumulative injection before pressure drive (0.14), original formation pressure (0.13), mid-depth of the reservoir (0.12), porosity (0.08), and crude oil (0.01) (Figure 14d. Correlation Coefficients of Factors Influencing Cumulative Oil Increment in the Mature Area Contributions).

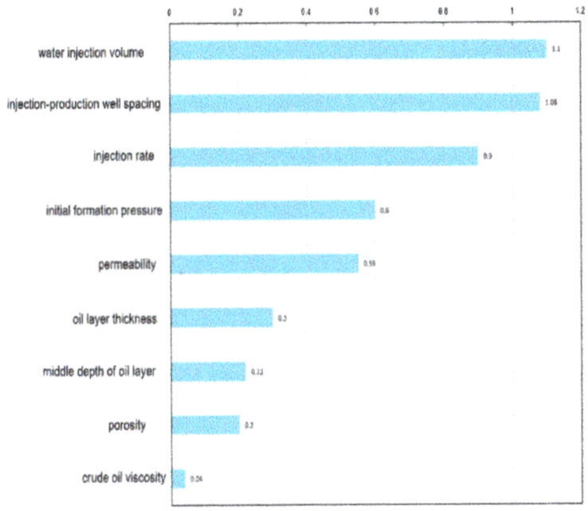

(**a**)

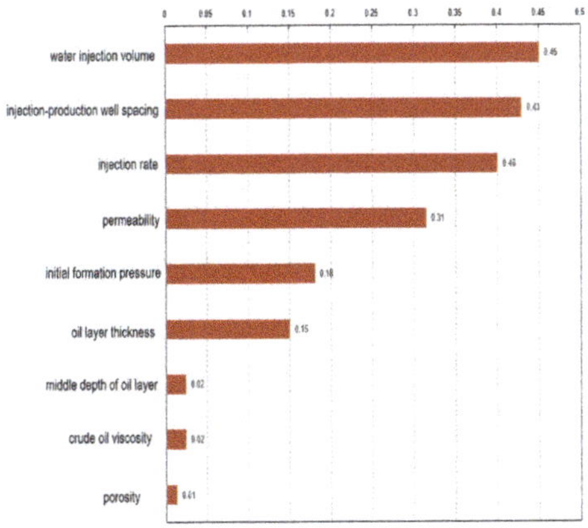

(**b**)

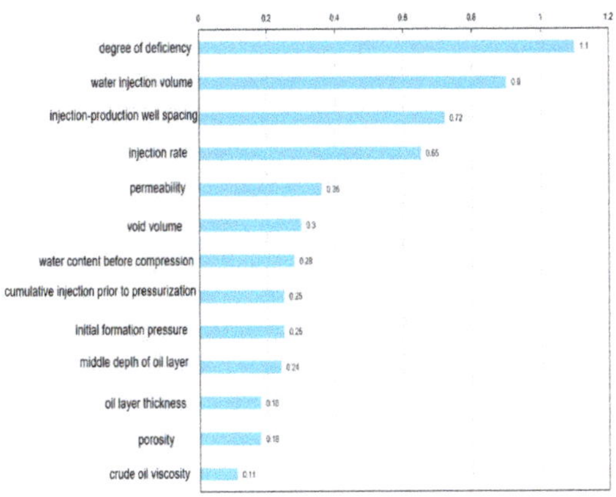

(**c**)

Figure 14. *Cont.*

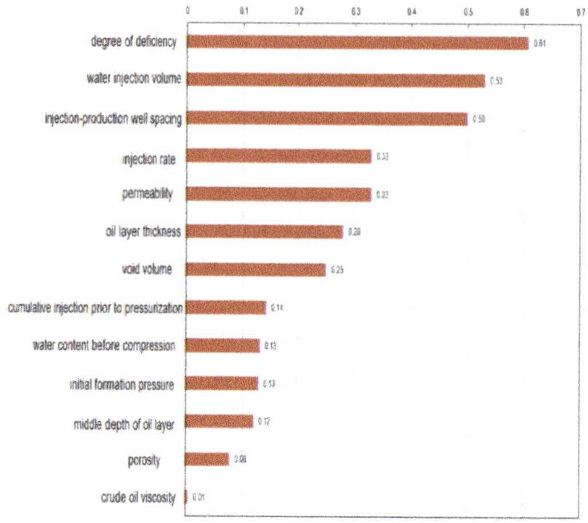

(**d**)

Figure 14. Evaluation of main controlling factors of pressure-driven technology in new and old areas. New area: pressure-driven water injection volume > injection–production well spacing > injection rate; Old area: degree of depletion > pressure-driven water injection volume > injection–production well spacing (data bar chart). (**a**) SHAP Values for Factors Influencing the Effectiveness of the New Area. (**b**) Correlation Coefficients of Factors Influencing Cumulative Oil Increment in the New Area. (**c**) SHAP Values for Factors Influencing the Effectiveness of the Mature Area. (**d**) Correlation Coefficients of Factors Influencing Cumulative Oil Increment in the Mature Area Contributions.

Based on the SHAP values of the influencing factors of the effectiveness status in the old area and the correlation coefficient values of the influencing factors of cumulative oil increase in the old area, the impact values of the main controlling factors of pressure-driven technology in the old area, from largest to smallest, are: degree of depletion > pressure-driven water injection volume > injection–production well spacing.

Oil wells with large depletions are more prone to water breakthrough, and the pressure-driven design in old fields should pay more attention to the imbalance of pressure and flow field caused by historical exploitation. Taking the well group C-1-1 as an example, well C-1-2 in this group has a depletion of 12.1%, well C-1-3 has a depletion of 15.4%, well C-1-4 has a depletion of 7.1%, and well C-1-5 has a depletion of 10.1% (Figure 15). According to the monitoring of the water injection front of well C-1-1, water breakthrough occurred in well C-1-3, which has a larger depletion.

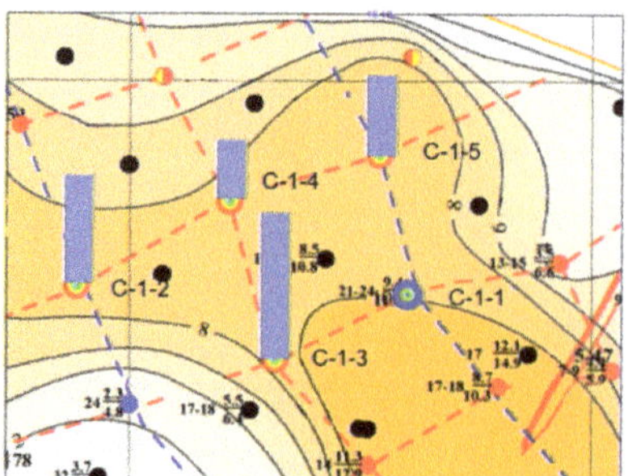

Figure 15. Bar Chart of Oil Well Depletion Degree in Well Group C-1-1.

Figure 16 illustrates the changes in the water injection front for Well Group C-1-1 at different time periods, specifically on 24 May 2023, 1 June 2023, and 12 June 2023. As shown in (Figure 16a), the water injection front change on 24 May 2023; in (Figure 16b), the water injection front change on 1 June 2023; and in (Figure 16c), the water injection front change on 12 June 2023.

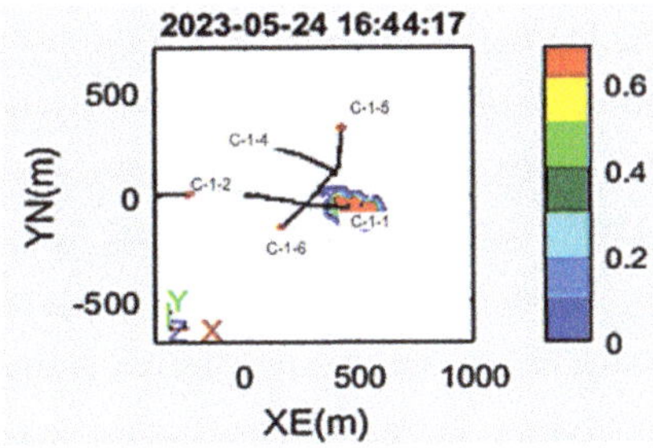

(a)

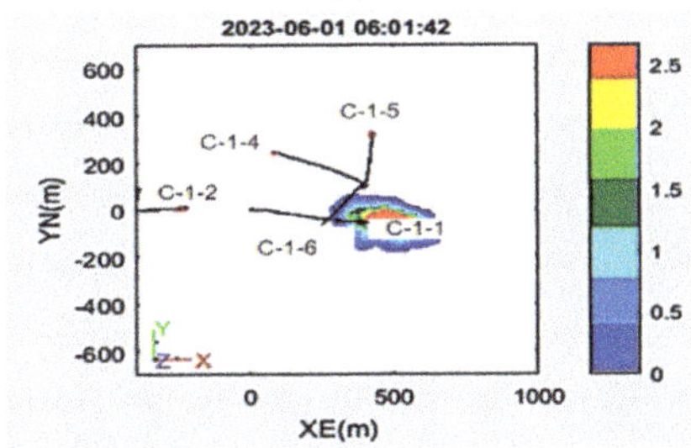

(b)

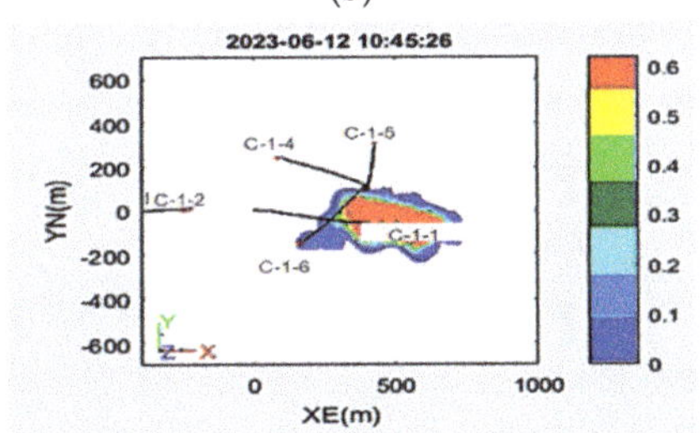

(c)

Figure 16. Variation in Water Injection Front at Different Time Periods in Well Group C-1-1. (**a**) Water Injection Front Variation Chart on 24 May 2023. (**b**) Water Injection Front Variation Chart on 1 June 2023. (**c**) Water Injection Front Variation Chart on 12 June 2023.

6. Prediction Model for Pressure Drive Effectiveness

A predictive model for pressure drive productivity, both non-time series- and time series-based, has been established utilizing data-driven approaches specific to pressure drive operations. This model aims to forecast the effectiveness of pressure drive measures by leveraging historical and real-time data pertaining to the process.

Flowchart for Establishing and Validating a Pressure-Driven Production Prediction Model with Non-Time-Series Data Sample Libraries, Including Data Preprocessing (Standardization), Selection of Appropriate Regression Algorithms (Figure 17).

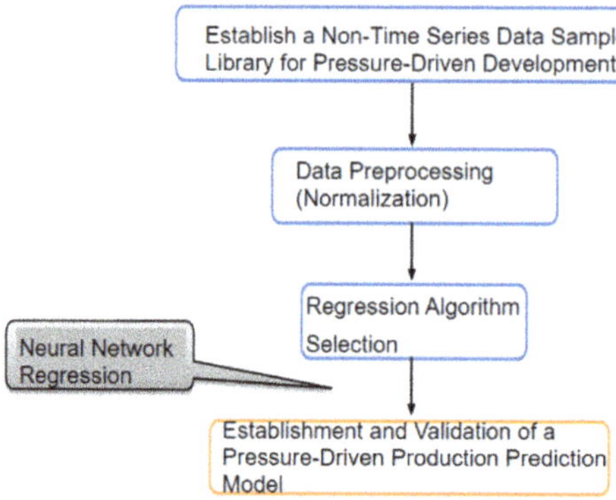

Figure 17. Flowchart of a Non-Temporal Pressure-Driven Production Prediction Model Based on Machine Learning.

In a neural network [41], multiple neurons can be combined to form a layer, and the stacking of these layers creates the network structure. We employ a Multi-Layer Perceptron (MLP).

$$z^{(l)} = f(W^{(l)}a^{(l-1)} + b^{(l)}) \tag{15}$$

where:

$z^{(l)}$ represents the activation output of the l \$-th layer.

$W^{(l)}$ denotes the weight matrix associated with the l \$-th layer.

$a^{(l)}$ is the activation output of the l-1 layer, which also serves as the input to the l layer.

$b^{(l)}$ is the bias vector for the l layer.

Non-time-series pressure-driven production capacity prediction model based on machine learning (neural network regression algorithm), with the neural network fitting diagram for the three months following pressure drive shown above (Figure 18).

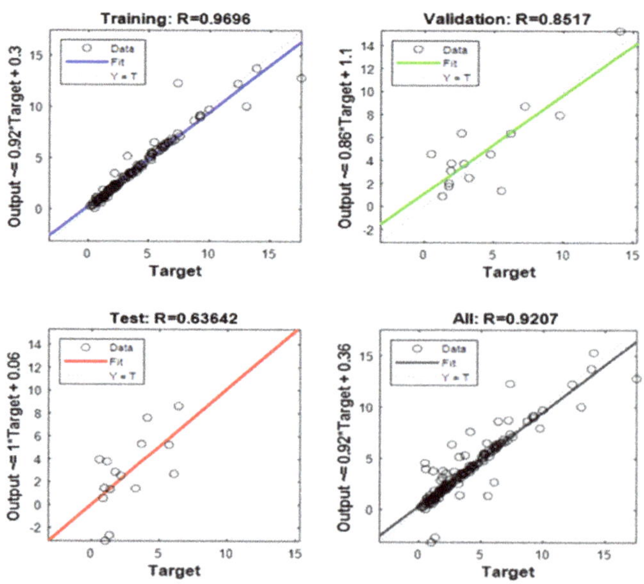

Figure 18. Neural Network Fitting Chart for Three Months After Pressure-Driven Injection.

Time Series Production Forecast Model Based on Deep Learning:

Based on the established pressure-driven production capacity prediction model, a regression plot for the daily oil production over three months was obtained, as shown in the above figure (Figure 19). A time series production forecast model employing a sequence-to-sequence (seq-seq) deep learning architecture has been developed. This model comprehensively considers not only the temporal influence of production dynamics on output but also the spatial impact of static geological parameters. The flowchart illustrating the model's framework is presented below (Figure 20). Results from both training and testing phases (Figure 21) demonstrate that the proposed deep learning-based time series production forecast model achieves high accuracy, underscoring its effectiveness in predicting production outcomes.

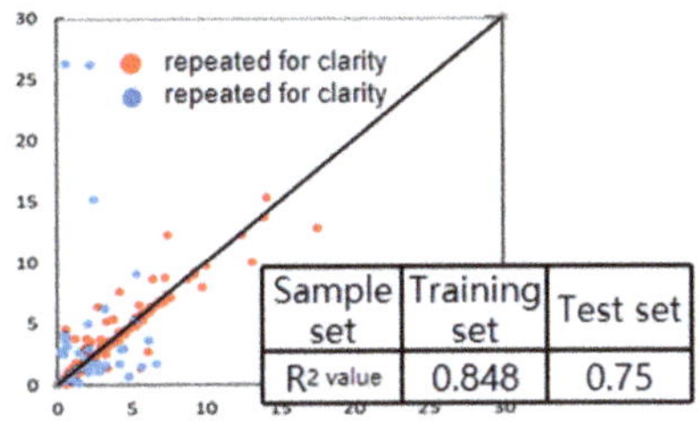

Figure 19. Regression Chart of Daily Oil Production Model for Three Months.

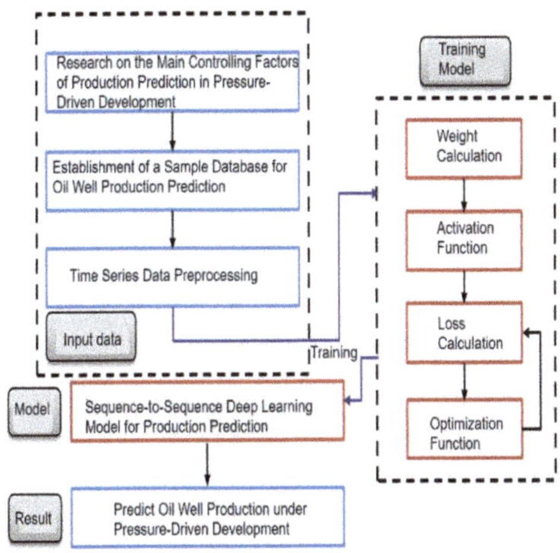

Figure 20. Time Series Production Prediction Model Based on Deep Learning.

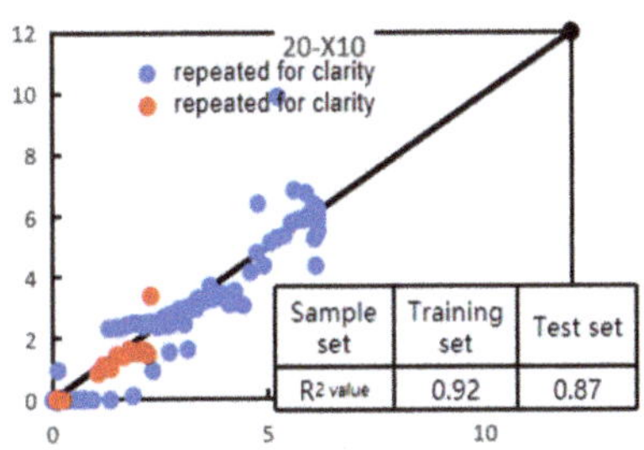

Figure 21. Time Series Production Regression Chart for Deep Learning.

Time series models for addressing pressure-driven problems can take various forms. We base our approach on the linear autoregressive model (AR) and incorporate pressure as an exogenous input variable. Let Pt denote the pressure at time t, and Xt represent the target variable at time t (e.g., flow rate, head height, etc.) The pressure-driven time series model can be expressed as:

$$X_t = c + \sum_{i=1}^{p} \phi_i X_{t-i} + \theta P_t + \varepsilon_t \tag{16}$$

where:

c is the constant term.
ϕi is the coefficient of the autoregressive term, indicating the influence of past values of the target variable on the current value.
p is the order of the autoregressive model.
θ is the coefficient of the pressure term, representing the impact of pressure on the target variable.
Pt is the pressure at the current time.
εt is the error term, assumed to be a white noise process with a mean of zero.

This model assumes that the target variable Xt is influenced by its own past values as well as the current pressure Pt. Such a model can be utilized to predict future values of the target variable (including production) given current and past pressure data.

7. Conclusions

This study delves into the exploitation mechanism of pressure drive by establishing mathematical models for oil-water seepage in tight oil reservoirs, a model of fracture permeability, and a prediction model for the pressure drive effect, elucidating the underlying mechanisms of pressure drive technology in enhancing oil recovery (EOR).

(1) An innovative approach to complex fracture networks was developed by investigating the strategy of pressure drive prior to fracturing, revealing that establishing a stable and continuous pressure profile with a high pressure coefficient is crucial for achieving long-term stable production, and also facilitates the formation of complex and branched fracture networks.
(2) Based on the varying physical properties of reservoirs, we innovatively developed a three-dimensional pressure drive method, optimizing different injection rates to control the propagation patterns of pressure-driven fractures, ultimately leading to high-pressure leak-off and the creation of complex fracture networks.
(3) The analysis identifies reservoir physical properties and injection displacement as the main controlling factors affecting the morphology and extension range of the fracture network.
(4) We conducted field-scale studies on the production characteristics of pressure drive operations, utilizing big data analytic tools such as SHAP analysis and correlation analysis to retrospectively evaluate the primary controlling factors of pressure drive in both new and established fields.

Compared to previous research on oil recovery solutions, the pressure drive method investigated in this study yields more pronounced results. Looking forward, we aim to explore deeper issues, expand our sample range, and experiment with different approaches to further advance our understanding and application of this technology.

Author Contributions: Conceptualization, H.L.; Methodology, B.J.; Validation, B.J.; Formal analysis, H.L.; Investigation, B.J.; Resources, B.J.; Writing—original draft, H.L. All authors have read and agreed to the published version of the manuscript.

Funding: This research was funded by ["Key Technology for Profitable Development of Untapped Reserves in Ultra-Low Permeability Oil Reservoirs" project by Sinopec] grant number [P24021] And The APC was funded by ["Key Technology for Profitable Development of Untapped Reserves in Ultra-Low Permeability Oil Reservoirs" project by Sinopec].

Data Availability Statement: The original contributions presented in the study are included in the article, further inquiries can be directed to the corresponding authors.

Conflicts of Interest: Author Haicheng Liu was employed by the Exploration and Development Research Institute of Sinopec Shengli Oilfield Company. The remaining authors declare that the research was conducted in the absence of any commercial or financial relationships that could be construed as a potential conflict of interest.

References

1. Qing, L.; Yuan, C.; Qiang, L. Establishment and evaluation of strength criterion for clayey silt hydrate-bearing sediments. *Energy Sources Part A Recovery Util. Environ. Eff.* **2018**, *40*, 742–750.
2. Li, Q.; Wang, Y.; Wang, F. Effect of thickener and reservoir parameters on the filtration property of CO_2 fracturing fluid. *Energy Sources Part A Recovery Util. Environ. Eff.* **2019**, *42*, 1705–1715. [CrossRef]
3. Guo, H.; Cheng, L.; Wang, P.; Jia, P. Water flooding experiment and law of carbonate reservoir cores with different fracture occurrences. *Pet. Geol. Recovery Effic.* **2022**, *29*, 105–112.
4. Yi, Z.; Yong, Y.; Zhi, S. Physical simulation of fracturing-flooding and quantitative characterization of fractures in low-permeability oil reservoirs. *Pet. Geol. Recovery Effic.* **2022**, *29*, 143–149.
5. Li, C. *Principles of Reservoir Engineering*; Petroleum Industry Press: Beijing, China, 2011; pp. 190–197.
6. Feng, Z. The Mechanism and Parameters-Optimization of Injection Fracture Pressure Propagation for Low Permeability Oil Field. Ph.D. Thesis, China University of Petroleum, Qingdao, China, 2009.
7. Jie, C.; Bin, J.; Jia, W. A novel numerical simulation of CO_2 immiscible flooding coupled with viscosity and starting pressure gradient modeling in ultra-low permeability reservoir. *Front. Earth Sci.* **2023**, *17*, 884–898.
8. Jie, C.; Jia, W. A new calculation method on the critical well spacing of CO_2 miscible flooding in ultra-low permeability reservoirs. *J. Porous Media* **2021**, *24*, 59–79.
9. Guan, Z.; Xiao, X.; Qing, Z. Study on hydraulic fracture propagation mechanism of pore-elastic media based on time-step linear superposition method. *Drill. Prod. Technol.* **2021**, *44*, 88–92.
10. Joseph, C.; Figueroa, A.A.; Ignacio, L.J. Self Adjusting Pressure Drive Shaft. Patent US20030226874A1, 11 December 2002.
11. Stines, J.R. Rotational Pressure Drive for a Medical Syringe. Patent US07/960022, 1 March 1994.
12. Pu, C.; Liu, Z.; Pu, G. On the Factors of Impact Pressure in Supercritical CO_2 Phase-Transition Blasting—A Numerical Study. *Energies* **2022**, *15*, 8599. [CrossRef]
13. Chang, X.; Yong, Z. Analysis of Oil Production and Fracturing Technologies for Low-Permeability Reservoirs. *Petrochem. Technol.* **2024**, *31*, 100–102.
14. Ze, R. High-efficient Development of Low-permeability Reservoir with Geology and Engineering Integration: A Case Study of N25 Block in Dongying Sag. *Drill. Prod. Technol.* **2024**, *47*, 90–96.
15. Shu, L.; Ding, Z.; Yong, W. Experimental Study on Micro-Nano Emulsion Injection Enhancement Technology for Low-Permeability Reservoirs. *Petrochem. Technol.* **2024**, *53*, 553–558.
16. Duo, P. Research on Enhanced Oil Recovery Technologies for Low-Permeability Reservoirs. *Petrochem. Technol.* **2023**, *30*, 50–52.
17. Liu, J.; Yao, Y.B.; Elsworth, D. Morphological complexity and azimuthal disorder of evolving pore space in low-maturity oil shale during in-situ thermal upgrading and impacts on permeability. *Pet. Sci.* **2024**, *1*, 1–20. [CrossRef]
18. Jin, Y. Fracturing Technology and Development Trends in Low-Permeability Oilfields. *China Pet. Chem. Stand. Qual.* **2023**, *43*, 175–177.
19. Yong, Y.; Shi, Z.; Xiao, C. Practice and understanding of pressure drive development technology for low-permeability reservoirs in Shengli Oilfield. *Pet. Geol. Recovery Effic.* **2023**, *30*, 61–71.
20. Bai, X.F.; Li, J.H.; Liu, X.; Wang, R.; Ma, S.; Yang, F.; Li, X.; Liu, J. Evolution of the Anisotropic Thermo-physical Performance for LowMaturity Oil Shales at an Elevated Temperature and Its Implications for Re-storing Oil Development. *Energy Fuels* **2024**, *38*, 15216–15224. [CrossRef]
21. Zhang, S.Q.; Liu, J.; Li, L.; Kassabi, N.; Hamdi, E. Petrophysical and Geochemical Investigation-Based Methodology for Analysis of the Multilithology of the Permian Longtan Formation in Southeastern Sichuan Basin, SW China. *Energies* **2024**, *17*, 766. [CrossRef]
22. Li, L.; Ma, S.; Liu, X.; Liu, J.; Lu, Y.; Zhao, P.; Kassabi, N.; Elsworth, D. Coal measure gas resources matter in China: Review, challenges, and perspective. *Phys. Fluids* **2024**, *36*, 071301. [CrossRef]
23. Chuan, C.; Huai, L.; Zhong, W. Analysis of pressures in water injection wells considering fracture influence induced by pressure-drive water injection. *Pet. Reserv. Eval. Dev.* **2023**, *13*, 686–694.
24. Yi, L.; Feng, W.; Yu, W. Mechanism of Enhanced Oil Recovery by Pressure Driving in Medium-Low Permeability Reservoirs. *Pet. Explor. Dev.* **2022**, *49*, 752–759.

25. Xiao, C. Practice and Understanding of Pressure-Driven Water Injection Technology in Binnan. *Inn. Mong. Petrochem. Ind.* **2024**, *50*, 43–46.
26. Feng, W.; He, X.; Yi, L. Mechanism of Enhanced Oil Recovery by Pressure Driving Technology with High Pressure Drop Adsorption. *Acta Pet. Sin.* **2024**, *45*, 403–411.
27. Qi, H.; Guo, P.; Yan, L. Pressure-Driven Water Injection Technology and Development Directions for Low-Permeability Reservoirs. *Contemp. Petrochem.* **2023**, *31*, 26–29.
28. Liu, Y.; Yamada, Y. The Characteristics and Design of Hybrid Pressure Drive System with Turbofan and Switch Valve for Laminated Foam-based Soft Actuator. In Proceedings of the JSME annual Conference on Robotics and Mechatronics (Robomec), Osaka, Japan, 6–8 June 2021; pp. 2A1–D02. [CrossRef]
29. Chiodini, G.; Paonita, A.; Aiuppa, A. Magmas near the critical degassing pressure drive volcanic unrest towards a critical state. *Nat. Commun.* **2016**, *7*, 13712. [CrossRef] [PubMed]
30. Velarde, E.; Ezcurra, E.; Horn, M.H. Warm oceanographic anomalies and fishing pressure drive seabird nesting north. *Sci. Adv.* **2015**, *1*, e1400210. [CrossRef]
31. Ishizuka, S.; Hamasaki, T.; Koumura, K. Measurements of flame speeds in combustible vortex rings: Validity of the back-pressure drive flame propagation mechanism. *Proc. Symp. Combust* **1999**, *27*, 727–734. [CrossRef]
32. Nakazawa, H.; Yokota, S.; Kita, Y. A Constant Pressure Drive System for Engine-Flywheel Hybrid Vehicles (First Report, Simulations for Fuel Saving Analysis of Hybrid Vehicles Using a Flywheel Pump/motor). *J. Jpn. Hydraul. Pneum. Soc.* **1996**, *27*, 513–518. [CrossRef]
33. Zhang, Z.; Xu, S.; Gou, Q.; Li, Q. Reservoir Characteristics and Resource Potential of Marine Shale in South China: A Review. *Energies* **2022**, *15*, 8696. [CrossRef]
34. Han, Z.; Dong, M.A.; Hou, B.W. Post-earthquake hydraulic analyses of urban water supply network based on pressure drive demand model. *Sci. Sin. Technol.* **2019**, *49*, 351–362. [CrossRef]
35. Jie, C.; JU, B.; Lyu, G. A computational method of critical well spacing of CO_2 miscible and immiscible concurrent flooding. *Pet. Explor. Dev. Online* **2017**, *44*, 815–823.
36. Jie, C.; Bin, J.; Wen, C. Laboratory simulation of CO_2 immiscible gas flooding and characterization of seepage resistance. *Front. Earth Sci.* **2023**, *17*, 797–817.
37. Jun, Z.; Peng, Y.; Su, W. 3D numerical simulation of fracture dynamic propagation in hydraulic fracturing of low-permeability reservoir. *Acta Pet. Sin.* **2010**, *31*, 119–123.
38. Shuai, T.; Yong, H.; Jie, Z. Reasonable pattern well spacing deployment of lens lithologic reservoirs with low permeability. *Lithol. Reserv.* **2018**, *30*, 116–123.
39. Ji, H. Application of advanced water injection inflow production relation model in low-permeability reservoir. *Spec. Oil Gas Reserv.* **2020**, *27*, 92–97.
40. Quan, Z.; Ming, L.; Zi, Z. Application of volume fracturing technology in tight oil reservoirs of Shengli oilfield. *China Pet. Explor.* **2019**, *24*, 233–240.
41. Song, F.; Ding, H.; Wang, Y.; Zhang, S.; Yu, J. A Well Production Prediction Method of Tight Reservoirs Based on a Hybrid Neural Network. *Energies* **2023**, *16*, 2904. [CrossRef]

energies

MDPI

Article

Dynamic CO_2 Leakage Risk Assessment of the First Chinese CCUS-EGR Pilot Project in the Maokou Carbonate Gas Reservoir in the Wolonghe Gas Field

Jingwen Xiao [1], Chengtao Wei [2,*], Dong Lin [1], Xiao Wu [1], Zexing Zhang [2] and Danqing Liu [2]

1 Research Institute of Safety, Environmental Protection and Technical Supervision, PetroChina Southwest Oil & Gas Field Company, Chengdu 610041, China; xiaojingwen@petrochina.com.cn (J.X.); lin_dong@petrochina.com.cn (D.L.); wuxiao1010@petrochina.com.cn (X.W.)
2 School of Environmental Studies, China University of Geosciences, Wuhan 430074, China; 17835033308@163.com (Z.Z.); liudq@cug.edu.cn (D.L.)
* Correspondence: chengtaowei@cug.edu.cn

Abstract: Existing CO_2 leakage risk assessment frameworks for CO_2 capture, geological storage and utilization (CCUS) projects face limitations due to subjective biases and poor adaptability to long-term scale sequestration. This study proposed a dynamic risk assessment method for CO_2 leakage based on a timeliness analysis of different leakage paths and accurate time-dependent numerical simulations, and it was applied to the first CO_2 enhanced gas recovery (CCUS-EGR) pilot project of China in the Maokou carbonate gas reservoir in the Wolonghe gas field. A 3D geological model of the Maokou gas reservoir was first developed and validated. The CO_2 leakage risk under different scenarios including wellbore failure, caprock fracturing, and new fracture activation were evaluated. The dynamic CO_2 leakage risk of the CCUS-EGR project was then quantified using the developed method and numerical simulations. The results revealed that the CO_2 leakage risk was observed to be the most pronounced when the caprock integrity was damaged by faults or geologic activities. This was followed by leakage caused by wellbore failures. However, fracture activation in the reservoir plays a neglected role in CO_2 leakage. The CO_2 leakage risk and critical risk factors dynamically change with time. In the short term (at 5 years), the project has a low risk of CO_2 leakage, and well stability and existing faults are the major risk factors. In the long term (at 30 years), special attention should be paid to the high permeable area due to its high CO_2 leakage risk. Factors affecting the spatial distribution of CO_2, such as the reservoir permeability and porosity, alternately dominate the leakage risk. This study established a method bridging gaps in the ability to accurately predict long-term CO_2 leakage risks and provides a valuable reference for the security implementation of other similar CCUS-EGR projects.

Keywords: CO_2 leakage; risk assessment; CCUS-EGR; numerical simulation; Wolonghe gas field

1. Introduction

CO_2 capture, geological storage and utilization (CCUS) is internationally recognized as one of the most effective means of mitigating global warming [1–3]. In general, CCUS is divided into three distinct components: CO_2 capture, utilization, and storage. CO_2 utilization is manifested mainly in CO_2-enhanced oil recovery (EOR) and CO_2-enhanced gas recovery (EGR) projects, photocatalytic conversion of CO_2 [4], and biological CO_2 conversion. CO_2 storage mainly includes geologic CO_2 storage (GCS), mineralization, and marine storage, among which GCS is the most important storage method. During the process of geological storage, CO_2 may leak through several paths, such as abandoned wells, reactivated faults and caprocks [5]. The leakage of CO_2 will pollute the groundwater and subsurface soil and also has a chance of being released into the atmosphere [6–8]. Before implementing any CCUS project, conducting a CO_2 leakage risk assessment should be a top priority [9].

Several CO_2 leakage risk assessment methods were proposed recently. The available methods can be divided into qualitative and quantitative assessment methods. Qualitative risk assessment methods primarily include the following frameworks: The Containment Assurance Scenario and Significance Identification Framework (CASSIF), developed by Ferhat Yavuz et al. [10], which employs fault leakage analysis, wellbore integrity evaluation, and seal failure detection to identify critical events in geological containment systems. This framework optimizes utilization of the Features, Events, and Processes (FEP) database through real-time scenario matching, significantly enhancing the modeling accuracy of risk significance identification. The Structured What-If Technique (SWIFT) proves particularly effective for assessing risks in inaccessible hazardous environments where consequence pathways remain ambiguous. Through systematic generation of hypothetical failure scenarios, SWIFT enables preemptive risk evaluation of complex industrial operations [11]. The Vulnerability Evaluation Framework (VEF) serves as a conceptual tool for systemic vulnerability assessment. This framework assists regulatory bodies and engineering professionals in establishing customized risk metrics while providing methodological guidance for design validation, risk quantification, and regulatory compliance management [12]. Earlier CCUS leakage risk evaluations mostly used evaluation frameworks and qualitative methods due to a lack of specific data, which resulted in greater uncertainty and risk misclassification. They could not dynamically assess the real-time changes in leakage risk during the sequestration process, and had limitations in decision-making. Thus, this method is gradually being replaced by quantitative risk assessment methods [11].

Quantitative risk assessment methods effectively make up for the shortcomings of qualitative assessment methods, mainly for the realization of quantitative portrayal of the risk of leakage, such as calculating the leakage volume or the risk of leakage from a single leakage path over a certain period of time [13]. They can accurately provide input for relevant monitoring and protection measures [14]. The main quantitative risk assessment methods include the overall assessment of CO_2 leakage risk with the Performance and Risk (P&R) [15] and the main potential leakage path identification methods are the Certification Framework Approach (CFA) [16], CO_2-Predicting Engineered Natural Systems (CO_2-PENS) [17], and Risk Interference Subsurface CO_2 Storage (RISCS) [18]. The main methods used recently for assessing the risk of CO_2 leakage in CCUS-EGR projects are described in the following section, but these various methods all have certain limitations. The experimental approach includes core replacement experiments, fracture expansion and permeability tests, etc. This method offers a visual representation of the reservoir's response to CO_2 injection, but its limitations are also evident. It cannot reflect the complex geological conditions of the actual subsurface reservoir, nor can it portray the long-term evolution of the geological sealing process. The Probabilistic Risk Assessment Method (PRA) is a statistical-based approach that utilizes leakage probability statistics and Fault Tree Analysis (FTA) to quantitatively assess the risk of leakage. However, the method is inherently dependent on data, and the lack of disclosure of leakage data from the CCUS-EGR project restricts the reliability of the results due to governmental policy. Machine learning methods, as one of the key quantitative risk assessment approaches, can use neural networks and other artificial intelligence algorithms to analyze field monitoring data from actual projects and predict leakage risk in conjunction with existing historical processes.

The primary benefit of this approach is the enhanced accuracy of the calculated leakage risk probability when combined with extensive data training calculations. However, the key disadvantage is the lack of interpretability, resulting in a black-box model that hinders the provision of reliable explanations for physical mechanisms [19]. Due to the lack of sufficient evaluation data and the dynamic nature of leakage risk over time, the accuracy and timeliness of existing quantitative evaluation methods are limited.

Numerical simulation is a mature and widely used tool for data generation [20]. It uses professional numerical simulation software, such as TOUGH (Lawrence Berkeley National Laboratory, Berkeley, CA, USA), CMG (Computer Modeling Group Ltd., Calgary, AB, Canada), ECLIPSE (Schlumberger Ltd., Houston, TX, USA), etc., to construct the corresponding numerical model and accurately predict the CO_2 migration distribution and leakage risk during the process of storage [21]. The integration of risk assessment and numerical simulation holds significant potential in addressing the insufficient dynamic assessment capabilities inherent in existing quantitative evaluation methodologies.

In recent years, quantitative evaluation methods have gradually become the mainstream method of risk assessment. Aiming at the problem of wellbore leakage in the process of CO_2 geological storage in a saline aquifer, Gao et al. [22] established a quantitative risk evaluation index system based on an analytic hierarchy process so as to determine the influence weight of each influencing factor on leakage according to the score. Sorgi et al. [23] proposed a quantitative assessment method for leakage risk, which introduced measurement, monitoring and verification methods to avoid qualitative analysis or subjective judgment involved in risk assessment and analysis. Unfortunately, the method lacks actual site applicability verification. Using a quantitative analysis method, Dou et al. [24] established a heat and mass transfer and distribution model of offshore gas production wells, explored the physical process of gas leakage and migration, quantitatively evaluated the influence of different factors, and designed a series of risk prevention measures. By comprehensively comparing qualitative, semi-quantitative and quantitative methods, the qualitative evaluation method has the advantages of simple operation and short completion time, but it has poor ability in later leakage quantification and risk assessment. The semi-quantitative method can achieve advantages in operability and flexibility; the advantage of the quantitative method is that it has sufficient data persuasiveness in terms of data accuracy and uncertainty processing. The disadvantage is that the operation is difficult and the completion time is long [25]. The quantitative evaluation method established in this paper effectively solves the problem of difficult operation. Users can assign values in turn according to the scoring table and combine the existing numerical simulation results of site injection, which greatly reduces the working time.

Internationally, Enhanced Gas Recovery (EGR) technology remains at a low technological maturity level, with limited documented implementation cases [26]. Notably, China achieved a milestone in 2024 by launching its first CCUS-EGR pilot demonstration project at the Wolonghe Gas Field. Characterized by advanced technical maturity and comprehensive supporting infrastructure, this pioneering project has successfully completed the first period of CO_2 injection.

Considering the limitations and shortcomings of the above risk assessment methods, this paper focused on identifying the leakage pathways, determining the values of various leakage indicators on the overall storage risk, and combining the results of numerical simulation to propose a leakage risk assessment method for long-term CCUS projects. The Maokou gas reservoir in the Wolonghe field, southwest China, was selected as an application case to investigate the dynamic assessment results of CO_2 storage risk under different time scales, which will provide a scientific method for field injection storage in the subsequent study area and a reliable basis for developing a leakage risk monitoring program.

2. Materials and Methods

2.1. Establishment of the Leakage Risk Assessment Method

2.1.1. Leakage Pathway Analysis

After CO_2 enters the formation through the wellbore, various factors during the injection or production may cause damage to the wellbore structure or lead to leakage along the fracture of an abandoned well. Then, the gas enters the reservoir and migrates upward due to buoyancy and accumulates under the caprock. Since the injection of CO_2 inevitably raises the pressure in both the reservoir and caprock, there is a possibility of fault or fracture activation, which in turn increases the risk of leakage [21]. With the increase of storage time, the dissolution of CO_2 in formation water and its water–rock reaction also poses a risk of changing the permeability of the reservoir's caprock [27]. Unexpected geological activities in the geological body will also damage the stability of the storage body and increase the risk of leakage. Therefore, in the above process, there will be a variety of potential leakage pathways of CO_2, mainly including the wellbore, caprock, reservoir, fault, fracture and so on (Figure 1a). Moreover, the risk of leakage through these pathways varies dynamically over time (Figure 1b). The temporal characteristics of these leakage pathways exhibit significant variability in their operational impacts. For instance, leakage through abandoned wellbores, pre-existing faults, or fractures—if it occurs—would likely manifest during the early storage phases and thus be readily detectable. In contrast, other latent risks, such as caprock integrity degradation induced by CO_2 injection-associated geomechanically stress redistribution and geochemical alteration processes, demonstrate progressive escalation with continuous CO_2 emplacement. These risks exhibit strong positive correlations with CO_2 concentration gradients. Consequently, the most effective dynamic risk quantification methodology involves integrating site-specific numerical simulations to systematically evaluate risk evolution through spatiotemporal analysis of CO_2 plume migration patterns.

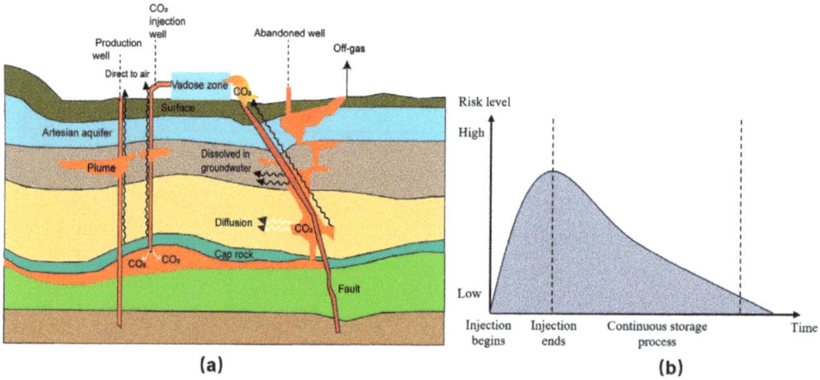

Figure 1. (**a**) Conceptual diagram of potential leakage path of CCUS, modified from Zhang et al. [28]. (**b**) Conceptual diagram of the risk of CO_2 sequestration and leakage in geological media, modified according to Gholami R et al. [29].

2.1.2. Evaluation Factors Determination

Based on the leakage pathways analysis listed above, the evaluation factors were divided into four categories: caprock, reservoir, wellbore, and geologic body safety and stability. The various evaluation factors can be further subdivided into more specific sub-factors, whose determination methods and scoring rules are shown below:

Caprock: The complete and continuous distribution of the caprock will affect the gas sequestration effectiveness in different areas of the site [30]. The thickness of the caprock can be divided into individual and cumulative caprock thickness. A greater thickness of the caprock generally correlates with enhanced safety. The permeability of the caprock directly influences the extent to which gas can escape after breaching the caprock. The lithology, mechanical stability, and chemical reactivity of the caprock mainly reflect its ability to resist external environmental changes and internal water–rock reactions over the long-term storage scale [31]. For long-term storage, it is also important to consider the content of non-reactive minerals in the caprock [32]. A series of sub-factors, including continuity, thickness, permeability, lithology, mechanical stability, and chemical reactivity, were established for accurate risk evaluation.

Reservoir: The influence of the reservoir on the overall closure performance is mainly reflected in the indirect effect on the upper caprock. The physical indices of the reservoir top adjacent to the caprock were selected to reflect this risk, in which the CO_2 mole fraction and the top pressure conditions can be used to visualize the strength of the effect on the caprock. Numerical simulations can be used to obtain the CO_2 mole fraction and top pressure distribution of the corresponding reservoir. For accurate risk evaluation, a series of sub-factors, including CO_2 mole fraction and the ratio of the top reservoir pressure to caprock rupture pressure, were established.

Wellbore: The main factors affecting the stability of a wellbore are completion time, abandonment time, cementing measures and historical accidents. The completion time determines the drilling technology and sealing methods used to complete the well [33]. The longer the abandoned time, the simpler the sealing measures taken at the time of abandonment. The quality of the cementing reflects the well's ability to withstand external interference over a long period [34]. The longer the cementing ring of the abandoned well, the better the safety of the well [5]. The last thing to emphasize is the history of technical accidents in the well. For accurate risk evaluation, a series of sub-factors were established, including the time of well completion, time of well abandonment, cementing related measures and historical engineering accidents.

Geological body safety: The main factors affecting the safety of the storage system include existing faults, potential rifts and seismic risk. As the dominant leakage path, a fault will greatly increase the risk of sealing leakage [35,36]. The peak acceleration of ground vibration is the horizontal acceleration corresponding to the maximum value of the ground vibration acceleration response spectrum. A smaller value indicates better overall safety and stability of the storage [37]. For accurate risk evaluation, a series of sub-factors including existing faults, potential rifts and peak acceleration were established.

2.1.3. Scoring Rules of the Sub-Factors

The scoring rules of sub-factors follow the principle of uniformity and grading according to the degree of influence on the leakage of the safety risk of the storage body. The greater the influence, the lower the score, with a minimum of 0.2, and the smaller the influence, the higher the score, with a maximum of 1. After that, the score will be graded according to the degree of influence, including 0.8, 0.6, 0.4 and so on. The specific scores of the risk levels are shown in Table 1.

The continuity of the caprock can be roughly categorized into three levels [30]. The thickness of the single layer at the bottom of the caprock is classified into three levels. The cumulative thickness of the caprock is divided into three classes. The caprock permeability is divided into three grades. The lithology of the caprock is divided into four grades according to the mud content. The mechanical stability of the caprock is divided into four grades. The chemical reactivity of the caprock is reflected by the content of non-reactive mineral content, which is roughly divided into four grades [32].

Table 1. Scores in relation to risk level.

Score	Risk Level
0.2	high
0.4	relatively high
0.6	medium
0.8	low
1.0	free

The CO_2 mole fraction is divided into five grades. When the top reservoir pressure approaches the caprock rupture pressure, the risk of rupture increases, which is divided into four grades.

The well completion time can be categorized into three grades. Wellbore abandonment time is divided into three grades on the basis of different times. The quality of cementing is qualitatively divided into three grades [34]. Finally, the history of accidents in wellbore is divided into three categories.

The distribution of faults or fractures in the caprock and reservoir can be classified into four grades on the basis of their scale. The grading of peak ground shaking acceleration is divided into five grades according to the actual seismic regions in China.

2.1.4. Evaluation Parameter Obtaining and Scores Calculation

The parameters of the caprock category index must be obtained from the corresponding stratigraphic data of the study area. The parameters in the wellbore category index and geological body safety category index, such as completion time and peak ground shaking acceleration, must be combined with the actual engineering situation, which can be obtained by consulting relevant data. The CO_2 mole fraction and pressure at the top of the reservoir, as part of the reservoir category index, need to be obtained with numerical simulation. The formation rupture pressure can be calculated using the Eaton method [38] based on the reservoir formation pressure evaluation method (Equation (1)):

$$P_f = \frac{\mu}{1-\mu}(P_0 - P_P) + P_P \tag{1}$$

In the formula:

P_f is the formation rupture pressure, MPa;

μ is the Poisson's ratio of the stratum, dimensionless;

P_0 is the overlying stratum pressure, MPa;

P_p is the formation pore pressure, g/cm^3.

The caprock juxtaposition thickness by fault is used to quantitatively characterize the vertical development of faults (Equation (2)) [36]:

$$H_j = H_c - H_f \tag{2}$$

In the formula:

H_j is the caprock juxtaposition thickness, m;

H_c is the caprock thickness, m;

H_f is the fault vertical throw, m.

During the process of geological storage, once a certain pathway leaks, it will cause huge safety hazards to the overall storage unit and even cause a series of malignant continuous safety accidents due to gas leakage. Therefore, we must consider the risks of each index equally.

The cumulative multiplication method is then adopted to comprehensively evaluate the CO_2 leakage risks by multiplying all indicators' scores, emphasizing time-amplified leakage impacts on system integrity as illustrated in Equation (3). Lv et al. [6] believed that there were differences in the hazard levels caused by different leakage pathways, and the scoring method was optimized on the basis of referring to the results of hazard classification. The corresponding evaluation score index coefficients were formulated for different types of leakage routes: wellbore, fault and geological activity are the first-level hazard indicators, and all sub-indicators need to be squared after multiplying the scores; the caprock and reservoir are the secondary hazard indicators, and the corresponding sub-index index coefficient is 1, that is, it remains unchanged.

$$\mathrm{SF} = (S_W * S_G)^2 * S_C * S_R \quad (3)$$

In the formula:

SF is the safe factor, which is equal to the result of multiplying the scores of all indicators in the assignment table;

S_w is the cumulative multiplication of score values of all wellbore indexes, given by the assignment table;

S_G is the cumulative multiplication of score values of all geological body safety indexes, which combines caprock faults, reservoir faults, and peak ground acceleration;

S_C is the cumulative multiplication of score values of all caprock indexes, given by the assignment table;

S_R is the cumulative multiplication of score values of all reservoir indexes, given by the assignment table.

The calculated SF classifies the risk into low, medium or high leakage levels, as outlined in Table 2.

Table 2. CO_2 leakage risk classification.

CO_2 Leakage Safety Index Calculation Results	Risk Level
0.1–1	low
10^{-4}–0.1	medium
$<10^{-4}$	high

2.2. CO_2 Leakage Risk Assessment of the Wolonghe CCUS-EGR Project

2.2.1. Background

The Wolong gas field is located in the Chongqing Changshou District, where the exposed Jurassic sand mudstone strata of the Ziliujing Group Dongyuemiao section represent the most ancient stratification. The gas field was first put into trial production in January 1972, and as of February 2023, the Wolonghe gas field had a cumulative total of 129 completed wells, with a cumulative gas production of 359.07×10^8 m^3. Utilizing this gas field as a paradigm, it is proposed that CO_2 geological storage be employed in conjunction with enhanced oil and gas recovery (CCUS-EGR).

This project is the first natural gas purification plant and gas carbon capture industrialization project in China, and it is also the first demonstration project of injecting CO_2 for enhancing natural gas recovery. Its carbon capture device is the first low pressure, medium and low concentration carbon capture device in China. The project adopted the mode of CCUS-EGR, and the chemical absorption method was used to capture the CO_2 discharged from the exhaust gas treatment unit. It is anticipated that about 26,500 tons of CO_2 can be captured annually and the gas recovery can be enhanced by 10% via CO_2 injection.

2.2.2. Numerical Model Establishment and Parameter Obtaining

The numerical model of the W67 well zone located in the Maokou Formation of the Wolonghe gas reservoir was established as shown in Figure 2. The professional reservoir simulation software CMG was employed by adopting the deterministic modeling method and the actual parameters were used to construct the geological model, which includes the well coordinates and the level structure, fault, well slope, geological layering, porosity, permeability and gas saturation data. At the same time, the quality of the geological structure, fault distribution and drilling stratification were strictly controlled, and the planar grid was dissected with a size of 50 m × 50 m. The basic parameters of the model are shown in Table 3.

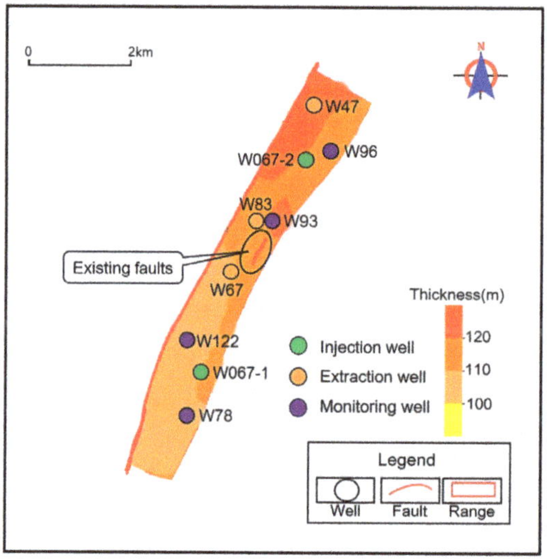

Figure 2. Thickness of caprock, well placement, and distribution of existing faults in the study area.

Table 3. Reservoir component model parameters and values.

Parameter	Value	Parameter	Value
surface area	19.7 km^2	permeability	0.1–134 mD
stratum	Maokou Group II	storage temperature	85.74 °C
central burial depth	3477 m	wellbore distribution	3 wells center,4 wells side
storage pressure	54.639 MPa	injection gas	Pure CO_2
gas reserves	66.2×10^8 m^3	number of wellbores	7
porosity	0~10%	number of grids	460,000

The caprock of the gas reservoir is the Longtan Formation, which consists of dense mudstone. The distribution is continuous, with a thickness of about 100~130 m. Its average permeability and porosity are 9.5×10^{-9} mD and 0.00385, respectively. The caprock does not contain reactive minerals and its mechanical properties are stable and not deformed.

The main component of reservoir gas is methane, accounting for 96.57%, while the remaining impurity gases are CO_2 and H_2S, accounting for 2.11% and 0.77%, respectively. The gas field is a dry gas reservoir characterized by ultra-low water saturation, with original connate water saturation ranging from 6.6% to 12.6%.

The permeability distribution of the numerical model relies on logging permeability data, and it was analyzed and calculated for different areas using well test data. The porosity distribution was derived from the logging porosity data and was analyzed using the sequential Gaussian simulation algorithm. The specific spatial distributions of the permeability and porosity are shown in Figure 3.

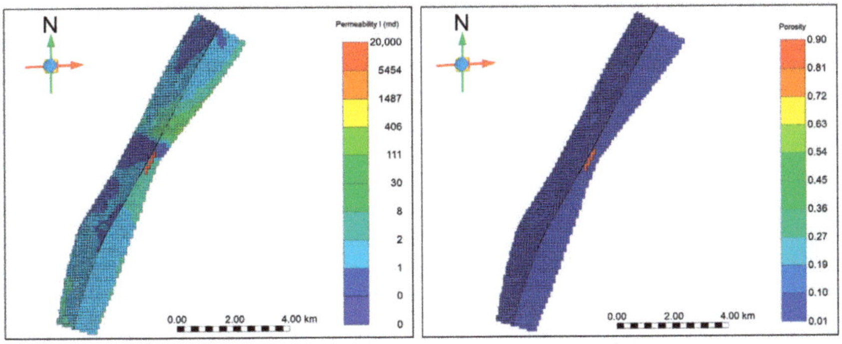

(a) (b)

Figure 3. Permeability (**a**) and porosity (**b**) distribution of the Wolonghe gas field W67 well zone.

There are two injection wells (W067-1 and W067-2) and three production wells (W47, W83, W67) located in the study area, as shown in Figure 2. The results of the cementing quality survey showed that there are 14 existing old wells, 10 wells that qualified for the pressure test, 2 wells that have no cementing quality test data and 2 wells that are unqualified (as shown in Table 4). All of the wells are now in a normal working condition, except for W122, which was abandoned in 2007.

Table 4. Wellbore information.

Well Number	Well Type	Establishment Time
W47	extraction well	1977
W83	extraction well	1985
W67	extraction well	1978
W93	monitoring well	1985
W122	monitoring well	2004
W78	monitoring well	1985
W067-1	injection well	2022
W067-2	injection well	2022

There is only one NE-SW fault in the gas production area of the W67 well, with a planar extension length of about 800 m. According to the relevant information, the seismic defense intensity of this site is 6 degrees, and the value of ground shaking acceleration is 0.05 g.

This CO_2-EGR project was planned into two phases. For phase I (2022–2027), a per-well CO_2 injection rate of 100,000 m^3/day and an aggregate production rate of 70,000 m^3/day for production wells were selected. During Phase II (2027–2053), representing the expansion phase, the injection rate was 150,000 m^3/day per injection well, while the total production rate was elevated to 110,000 m^3/day.

2.2.3. Leakage Scenarios Set up

Based on the identification and analysis of leakage paths, this paper established a leakage risk assessment framework. However, in the actual storage process, various unpredictable sudden leakage situations may occur. Therefore, it is necessary to select representative indicators in the evaluation framework and combine them with the actual engineering situation to predict possible leakage scenarios. This measure can provide a reliable basis for verifying the accuracy and applicability of the evaluation framework to a certain extent.

Building upon the risk assessment indicators and the actual conditions of the study area (as shown in Figure 2), three scenarios, namely, CO_2 leakage along the well, leakage when the cap layer is damaged by faults, and new faults generation in the reservoir were set up. The detailed parameter settings in different scenarios are listed in Table 5.

Table 5. Parameter settings for different simulation groups.

Division of Groups	Longitudinal Extension of the Fault to the Length of the Cap
Control group	No fault
Experimental group a	50 m
Experimental group b	100 m
Experimental group c	200 m

Leakage along the wellbore: In order to simulate leakage along the injection well, a specific grid near the injection well was set as a single fracture. The permeability of the fracture grid is 20,000 mD, the porosity is 0.9, and the rest of the parameters are consistent with the original model [39].

Leakage from new faults in the reservoir: The new fracture network at the edge of the gas production area was used to simulate the addition of new faults caused by geological activity. The permeability of the fracture network is 20,000 mD, the porosity is 0.9, and all other parameters are the same as in the original model.

Leakage of caprock damaged by faults: Three simulation groups (a, b, and c) were set up, with fault longitudinal extension thicknesses of 50 m, 100 m, and 200 m, respectively (Table 5 and Figure 4). These groups simulated different fault extension distances triggered by varying levels of geological activity. The permeability of the fracture grid is 20,000 mD, the porosity is 0.9, and all other parameters remain consistent with the original model.

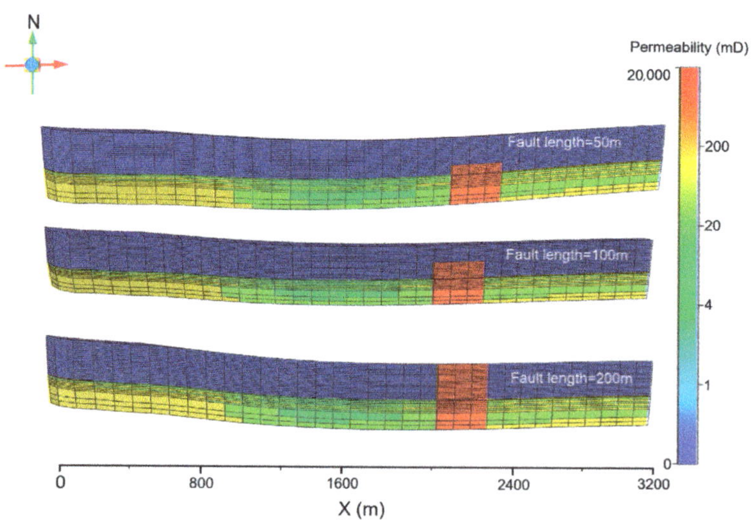

Figure 4. Longitudinal cross section of the caprock in different experimental groups (caprock layer in blue, reservoir in yellow–green, fault in red).

3. Results and Discussion

3.1. CO_2 Dynamic Leakage Risk Assessment Framework

Aiming at the identified factors and sub-factors, the grading assignment method was used to determine the corresponding scores of different levels according to the influence of different indicators on the leakage risk. The specific scoring rules for different risk factors are shown in Table 6.

Table 6. Sequestration system risk scoring of indicators.

Class I Indicators	Class II Indicators	Basis of Classification	Corresponding Score
Caprock	Caprock integrity continuity	Cover trap area > 80% and no broken area	1
		Cover trap area 40–80%	0.8
		Cover trap area < 40%	0.4
	Thickness of a single cap layer immediately adjacent to the reservoir	>20 m	1
		10–20 m	0.8
		<10 m	0.6
	Cumulative caprock thickness	>300 m	1
		150–300 m	0.8
		<150 m	0.6
	Caprock permeability	$<10^{-3}$ mD	1
		10^{-3}–10^{-2} mD	0.6
		$>10^{-2}$ mD	0.4
	Caprock rock type	Paste rock, mudstone, calcareous mudstone	1
		Sandy or silty mudstone	0.8
		Sandy mudstone, silty mudstone	0.6
		Shale, dense graywacke	0.4
	Caprock mechanical stability	Rock Young's modulus > 18 Gpa	1
		Rock Young's modulus 15–18 Gpa	0.8
		Rock Young's modulus 12–15 Gpa	0.6
		Rock Young's modulus < 12 Gpa	0.4
	Caprock non-reactive mineral content	>80%	1
		60–80%	0.8
		40–60%	0.6
		20–40%	0.4
		0–20%	0.2
Reservoir	The CO_2 mole fraction of top reservoir	0–10%	1–0.9
		10–30%	0.9–0.7
		30–50%	0.7–0.5
		50–70%	0.5–0.3
		70–100%	0.3–0
	Top reservoir pressure/caprock rupture pressure	0.4–0.5	1
		0.5–0.6	0.8
		0.6–0.8	0.6
		>0.8	0.4

Table 6. *Cont.*

Class I Indicators	Class II Indicators	Basis of Classification	Corresponding Score
Wellbore	Wellbore engineering project	Non-existent	1
		Existent	0.4
	Wellbore completion time	Completion time < 18 years	1
		Completion time > 18 years but completed after 1992	0.8
		Completed before 1992	0.4
	Wellbore abandonment time	Less than 5 years	1
		5–13 years	0.8
		More than 13 years	0.6
	Wellbore cementing quality	Excellent cementing quality	1
		Medium cementing quality, partially disconnected free well sections	0.8
		Poor cementing quality, generous connected free well sections	0.4
	Cementing depth of cemented ring	Sealed to surface	1
		Unsealed to surface	0.6
	Length of cement ring cementation for abandoned wellbore	>30 m	1
		5–30 m	0.8
		0–5 m	0.6
		No measures taken	0.2
	Wellbore historical engineering incidents	All free of engineering incidents	1
		Technical events such as overflows and well surges	0.6
		Had well blowout	0.2
Geological body safety	Fault or rift development in the caprock	No faults or fractures	1
		H_j > 100 m	0.8
		H_j 50–100 m	0.6
		H_j < 50 m	0.4
	Fault or rift development in the reservoir	No faults or fractures	1
		Limited faults and intact caprock	0.8
		5–10 large-scale faults (extending more than 1 km)	0.6
		More than 10 large-scale faults or have fracture zone (extending more than 1 km)	0.4
	Peak ground acceleration	<0.05 g	1
		0.05–0.1 g	0.8
		0.1–0.2 g	0.6
		0.2–0.3 g	0.4
		>0.3 g	0.2

3.2. Validation of the Numerical Model of the Wolonghe EGR Project

It is important to validate the established numerical model with historical production data to guarantee the accuracy and scientific validity of the simulation results. In this paper, the historical gas production rate of W47 and wellhead pressure were employed to calibrate the numerical model by model fitting, and the fitted results are shown in Figure 5.

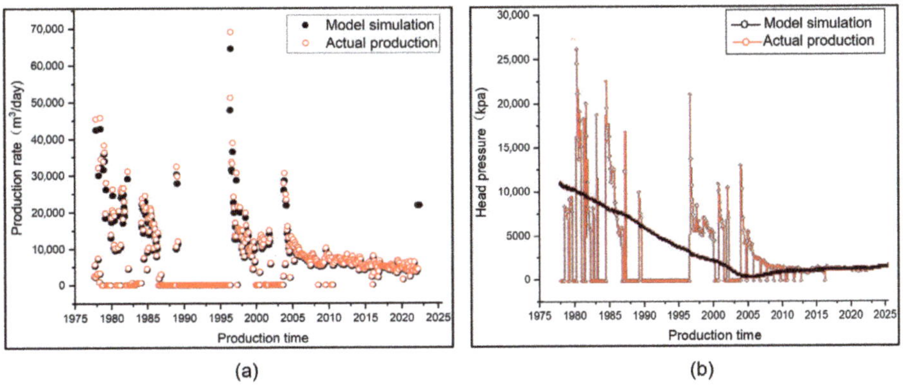

Figure 5. Model production rate and pressure fit for well W47. ((**a**) is W47 production rate validation results; (**b**) is W47 head pressure validation results).

The fitting results on gas production rate show good agreement with the historical data, especially after 2005. In the early stages of production, due to the lack of historical data, there is a large error between the simulated and actual values of wellhead pressure. However, with the enrichment of production data samples after 2010, the fitting results of wellhead pressure have become increasingly realistic. Overall, it can be seen that the numerical model constructed in this paper demonstrates a good historical fitting effect. The model is accurate and reliable, and the simulation results obtained can be used as an important basis for subsequent risk assessment and the development of monitoring programs.

3.3. Simulation Results Under Normal Operating Conditions

The simulation results of the CO_2 mole fraction distribution at the top of the reservoir for 5 years of gas injection, 10 years of gas injection, and 30 years of gas injection under normal operating conditions are shown in Figure 6.

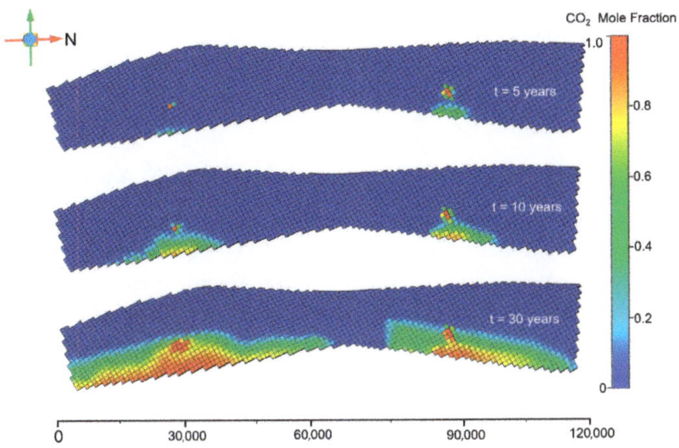

Figure 6. CO_2 distribution at the top of the reservoir at different gas injection times under normal operating conditions.

The permeability distribution of the reservoir exhibits a distinct pattern of higher values in the eastern region compared to the western region, consequently leading to preferential eastward migration of injected CO_2 (Figure 6). Figure 7 shows the CO_2 migration distance at different times of gas injection. For phase I (injection rate: 100,000 m^3/day, duration: 5 years), the analysis reveals that the maximum north–south migration distance of CO_2 from the southern injection well reaches approximately 10,000 m, while the northern injection well demonstrates a greater migration extent of 12,000 m along the same orientation. Additionally, a discernible migration plume extending approximately 8000 m from the injection wells toward the eastern boundary is observed. After 10 years of continuous injection, the migration distances exhibit a progressive enlargement. The southern injection well shows a significant increase in north–south migration distance, reaching 16,000 m, whereas the northern counterpart reaches 14,000 m. The CO_2 plume maintains its eastward migration tendency toward the boundary. Reaching 30 years of injection duration, both injection wells demonstrate extensive CO_2 migration. The maximum north–south migration distances escalate to 30,000 m (southern well) and 25,000 m (northern well), respectively. Notably, the eastward migration distance from injection wells to the eastern boundary attains 10,000 m. The overall spatial distribution consistently manifests higher migration potential in the eastern sector relative to the western region, aligning with the inherent permeability characteristics of the reservoir.

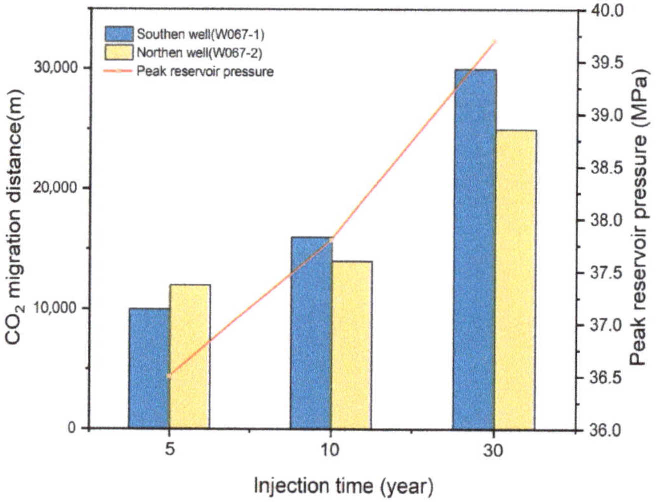

Figure 7. CO_2 migration distance and peak pressure of the reservoir at different times of gas injection.

Pressure variations induced by CO_2 injection constitute a critical monitoring parameter. Figure 7 shows the peak pressure curve and Figure 8 illustrates the spatiotemporal evolution of reservoir pressure under normal operating conditions across different injection durations. The spatial pressure distribution demonstrates a strong positive correlation with the CO_2 plume dispersion pattern. At the 5-year injection period, the maximum reservoir pressure reaches 36.5 MPa, escalating to 37.8 MPa after 10 years of continuous injection. Upon reaching 30 years of sustained injection, the formation pressure peaks at 39.7 MPa. Notably, short-term injection (<5 years) induces minimal pressure deviation from initial formation conditions, whereas the 30-year injection duration triggers a marked pressure surge in the eastern sector. This results in a distinct west-to-east pressure gradient, consistent with the reservoir's intrinsic permeability anisotropy.

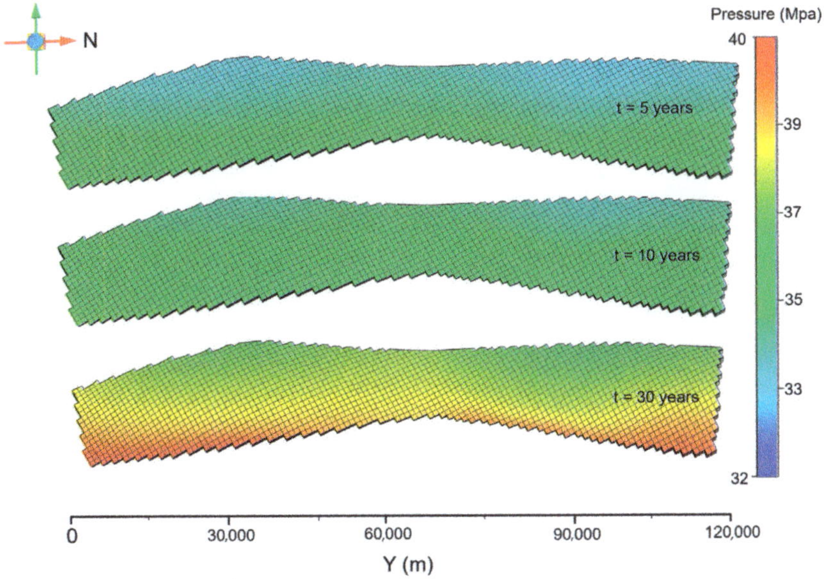

Figure 8. Pressure distribution at the top of the reservoir at different times of gas injection.

The reservoir exhibits ultra-low water saturation characteristics (ranging from 6.6% to 12.6%), significantly suppressing geochemical reaction intensities under current conditions. Because of the unique carbonate reservoir in the study area, numerical simulation was used to investigate the effect of the CO_2 water–rock reaction process on the amount of CO_2 leakage via the cap rock, as shown in Table 7.

Table 7. Comparative analysis of CO_2 leakage capacity under different scenarios (t).

	Water–Rock Reaction	No Water–Rock Reaction	Relative Change	Average Annual Change
10 years	1.01	1.14	0.13	0.013
20 years	9.68	10.56	0.88	0.044
30 years	32.65	35.33	2.68	0.089

From Table 7, it can be seen that the partial CO_2 leakage from the caprock after the addition of the water–rock reaction is slightly less than that from the cap without the addition of the water–rock reaction. Considering that some of the CO_2 may dissolve simultaneously, it is assumed that part of the CO_2 also passively participates in the mass consumption of the water–rock reaction process, which reduces the mass of gas diffused into the caprock, and the amount of CO_2 leakage from the caprock may be reduced. The average annual change in the leakage mass of the two is less than 0.1 t over a longer reaction time, which is a very small error compared to the uncertainty of the real situation over a long-time scale simulation. Therefore, it can be assumed that the CO_2–water–rock chemical reaction has a relatively small impact on the caprock. For this reason, water–rock interactions were excluded from the analytical framework due to their negligible thermodynamic impact.

3.4. CO_2 Leakage Under Different Leakage Scenarios

3.4.1. Spatial Distribution of CO_2 Under Different Scenarios

The CO_2 distribution at the top of the reservoir under the wellbore leakage scenario is shown in Figure 9. The maximum north–south migration distances show minimal variation from those under normal operating conditions; however, significant divergence is observed in both the maximum CO_2 mole fraction value and the spatial distribution patterns of high-concentration CO_2 accumulation zones during equivalent operational timeframes. According to the simulation results, 5 years after CO_2 injection begins, the CO_2 gas fraction in the fracture network of the reservoir appears to be significantly increased due to the presence of fractures in the injection well. The maximum mole fraction is only about 0.8. In contrast, there is no large-scale high-concentration CO_2 accumulation zones at the top of the reservoir. After 10 years of injection, the CO_2 shows a tendency to diffuse from the fracture to the surrounding area and the maximum fraction increases to 1.0. When the gas injection time reaches 30 years, the gas concentration in most of the grid appears to increase sharply, and the high gas concentration zone is concentrated near the two injection wells, showing a migration trend from a low concentration in the west to a high concentration in the east. The CO_2 migration area expands to an impressive 1100 km^2.

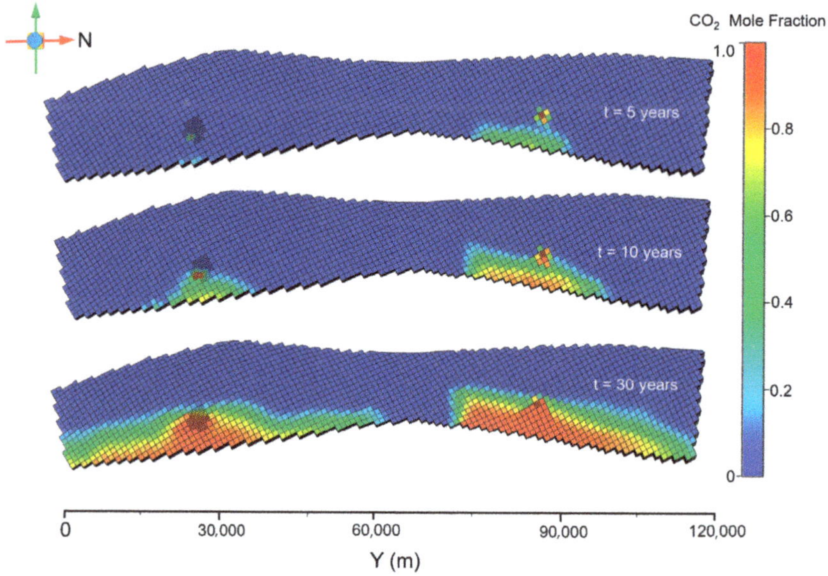

Figure 9. Wellbore leakage scenario CO_2 distribution at the top of reservoir for 5, 10 and 30 years of gas injection.

When compared to the no fracture control group, the presence of fractures in the injection well may reduce the concentration of gas components in the original migration grid. However, excessive gas concentrations concentrated near the fracture can lead to a greater susceptibility to excessive gas pressures at specific points, increasing the risk of caprock breach at particular points.

The CO_2 distribution at the top of the reservoir in the caprock leakage scenario is shown in Figure 10. The maximum north–south migration distances demonstrate minimal variation from those under normal operating conditions. By comparing the simulation results, it is found that the damage to the caprock caused by faults has less influence on the distribution of CO_2 gas in the reservoir. The spatial distribution of CO_2 is basically the same as the normal working condition.

The CO_2 distribution at the top of the reservoir in the fault leakage scenario is shown in Figure 11. After 5 years of gas injection, the total amount of injected gas is small, and the presence of new faults buffers the high-pressure area generated by the gas injection. The concentration of CO_2 components decreases significantly in the area close to the new fault, presumably because the presence of the fault promotes gas migration to the surrounding grids and the upper caprock. After 30 years of gas injection, the CO_2 content in the grid at the top of the reservoir near the injection well area increases significantly. The decrease in gas content in the fault grid indicates that this part of the gas has already diffused into the upper caprock, which increases the possibility of gas leakage from the cap layer.

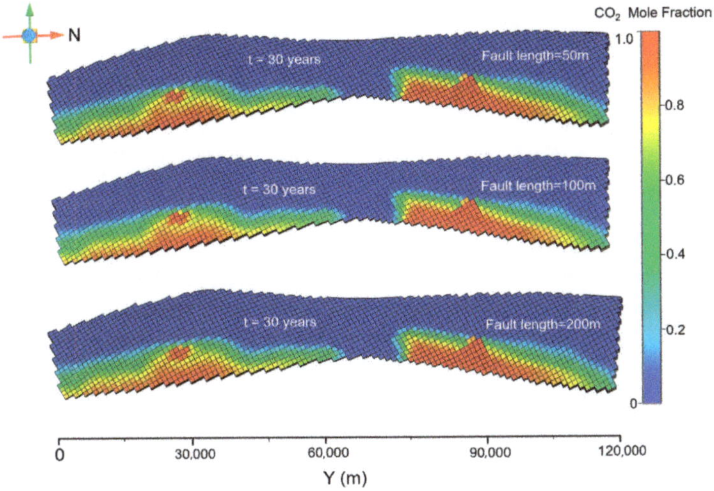

Figure 10. Caprock leakage scenario CO_2 distribution at the top of reservoir for 30 years of injection in different groups.

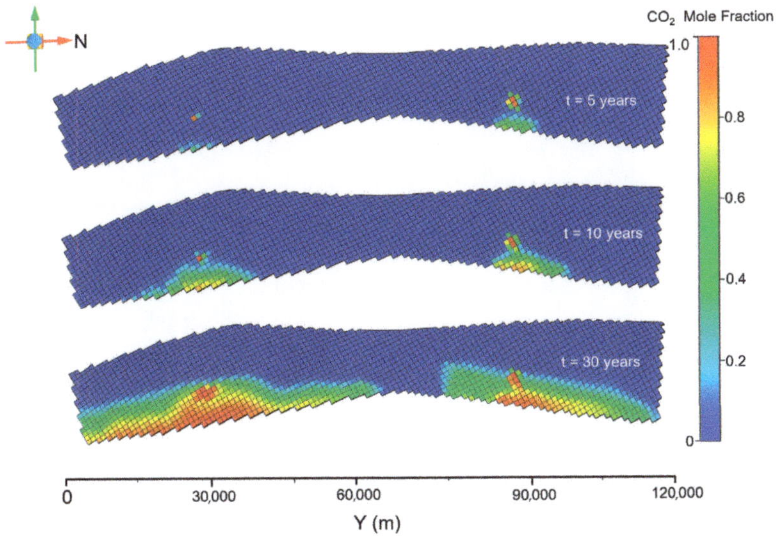

Figure 11. Fault leakage scenario CO_2 distribution at the top of reservoir for 5, 10 and 30 years of gas injection.

3.4.2. Leakage Capacity of CO_2 via the Caprock Under Different Leakage Scenarios

The results of the CO_2 leakage calculations for the caprock section are shown in the Table 8. The presence of fractures in the injection wells exacerbates the process of CO_2 leakage in the caprock formation, and the amount of leakage generated by a small volume of fractures is extremely large, far exceeding that of the control group by a factor of 600. Moreover, the presence of new faults does not have a significant effect on CO_2 leakage in the caprock. Even with an injection time of up to 30 years, the difference is very small. As long as reservoir faults do not disrupt the upper caprock, some smaller-scale additions to the reservoir area will not affect the caprock's sealing capacity.

Table 8. Caprock CO_2 leakage capacity (t) under different scenarios.

Injection Time	Control Group	Wellbore Fracture Group	Sim. a	Sim. b	Sim. c	New Fault Group
5 years	0	1635.7	110.03	178.64	296.75	0
10 years	0	4936.8	232.67	388.26	652.96	0
30 years	8.4	26,278.2	892.58	1591.83	1800.92	5.19

As shown in Figure 12, a more accurate quantitative assessment of the gas leakage is provided. During the 30 years injection period, the CO_2 leakage in the experimental group is up to 160–320 times higher than that of the control group. It can be seen that the gas leakage increases sharply as the injection time increases, and the length of the flaw extending into the caprock also affects the leakage amount. When the fault destroys the entire caprock, the integrity and sealing of the cap layer will no longer exist.

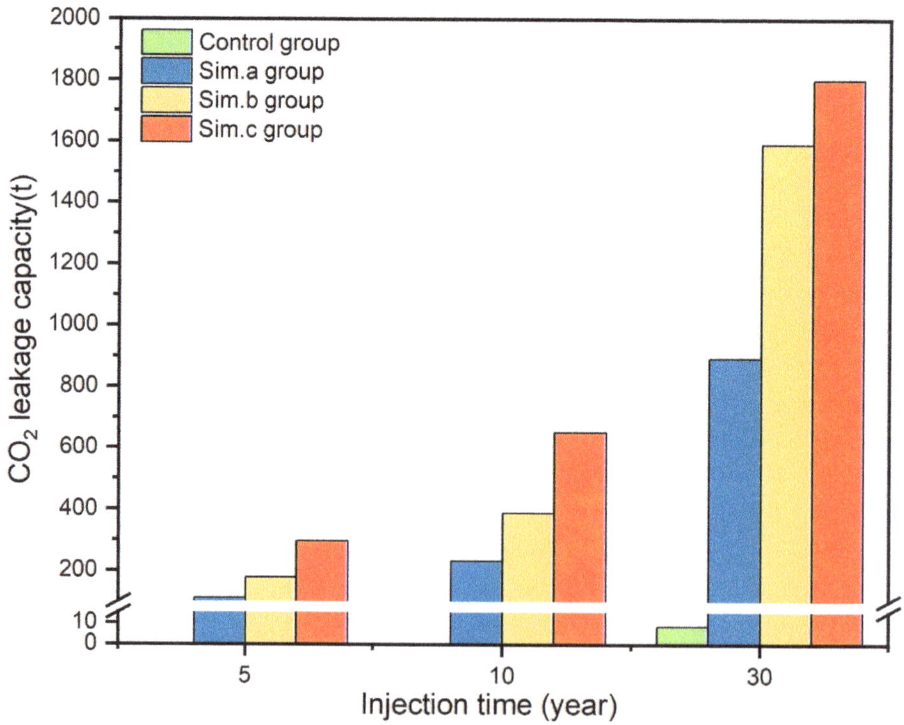

Figure 12. Comparison of CO_2 leakage from caprock under different leakage scenarios.

3.5. Sensitivity Analysis

The sensitivity analysis of the simulation in this paper is mainly divided into two categories: a sensitivity analysis of the actual site and a sensitivity analysis of the leakage path. For the verification of the actual site model, this paper fully investigated the geology data, constructed the regional scale model based on the actual stratum conditions, and completed accuracy verification of the model based on the historical production data. It can be considered that the model parameters are fully applicable to the actual site.

Secondly, this paper focused on the main leakage path in the geological storage process. After establishing the model, the representative leakage situation was selected for simulation prediction. The prediction results also further reflect that the caprock is the dominant factor determining storage risk. Other model parameters were less sensitive to the overall storage, which verifies that the evaluation scoring mechanism established in this paper is feasible.

3.6. Leakage Risk Assessment of the Wolonghe CO_2-EGR Project

On the basis of the different working stages of the injection and extraction wells in the actual sequestration process, the leakage risks at 5 years after the start of injection (2027) and 30 years after the expansion of the injection volume (2053) were calculated and the corresponding risk grading diagrams were drawn.

According to the leakage rating method in Table 6, a higher score indicates a safer containment process. Areas with a risk score greater than 0.1 have a low risk of leakage and are acceptable under normal working conditions. Conversely, areas with a risk score of less than 0.1 have a higher potential risk of leakage.

With an injection time of 5 years, the potential leakage risk is mainly located in the vicinity of the W83 and W93 wells, which were analyzed due to the poor cementing quality of the W83 wells, and there is a risk of leakage along the wellbore. Another potential risk point is near the W47 well, where poor cementing quality in the B and C annulus of the well increases the risk of gas leakage. In addition, the risk of leakage around the two injection wells is also high, and it is recommended that appropriate monitoring programs be implemented. Overall, the leakage risk is very low under the condition of 5 years of gas injection and the containment is in an ideal sealing condition. However, individual engineering wells may increase the leakage risk in the adjacent area due to the time of well formation and cementing quality. Therefore, it is necessary to strengthen monitoring efforts and take the necessary preventive measures.

In the second stage of extending injections to 2053, the points with potential leakage risk show a regionalized increase, mainly distributed on the eastern side of the study area. The analysis suggests that there are some areas of higher permeability on the east side of the reservoir, forming a dominant permeability path. The process of sequestration promotes further gas migration, which increases the risk of leakage when the caprock is breached. Overall, as storage time and the total amount of injected gas increases, CO_2 will eventually accumulate at the top of the reservoir due to its own buoyancy, significantly increasing the average molar fraction of CO_2 and the formation pressure. Referring to the scoring rules in Table 2, the overall score is significantly lower compared to 2027, which is consistent with the increased risk of leakage after actual injection. For long-term storage, the high-risk areas shown in Figure 13 need to be equipped with appropriate monitoring points to track the leakage risk in real time, ensuring that the dangerous situation at individual risk points does not deteriorate, which could lead to a dramatic spread of gas in the local area or even throughout the entire study area.

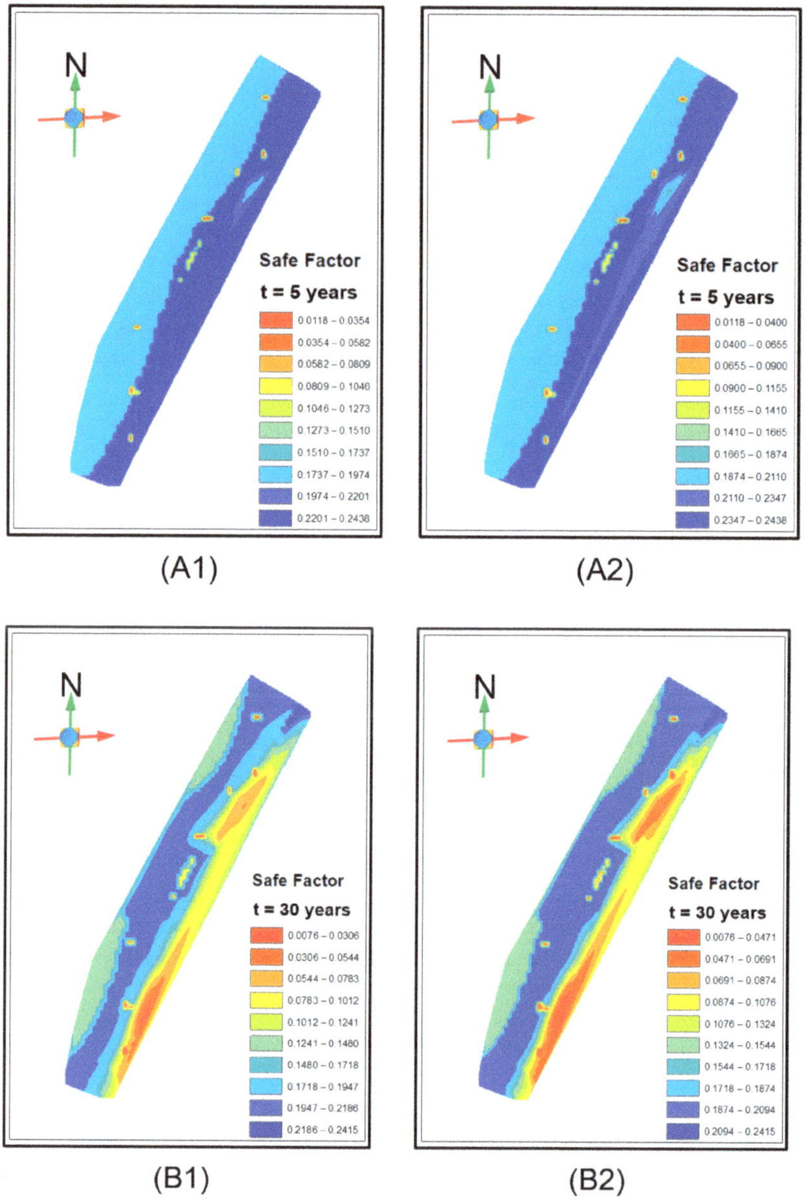

Figure 13. Map of safe factors for the study area at different injection times ((**A1**) is the safe factor spatial distribution at 5 years injection; (**B1**) is the safe factor spatial distribution at 30 years injection; (**A2**) and (**B2**) are the safe factor plots after superimposing the spatial distribution of CO_2 mole fraction).

4. Conclusions

This paper proposed a dynamic assessment method for CO_2 leakage risk assessment and applied it to the CO_2-EGR project in the Wolonghe gas field, which is the first CO_2-EGR pilot project in China. The conclusions are summarized as follows:

(1) Considering the necessity of constructing a numerical model for each CCUS project, a more precise dynamic evaluation method for assessing CO_2 leakage risk was developed. A comprehensive evaluation index and a scientific score calculation method were estab-

lished according to the characteristics of different leakage modes and their influencing factors.

(2) An accurate 3D numerical model of the CO_2-EGR project in the Maokou gas reservoir was developed and validated using detailed geological and hydrological data and historical production data. The CO_2 leakage risks under different scenarios were evaluated and the results indicated that the CO_2 leakage risk was observed to be most pronounced when the caprock integrity was damaged by faults or geologic activities. This was followed by leakage caused by wellbore failures.

(3) According to the dynamic CO_2 leakage risk assessment results, this project has a low risk of CO_2 leakage, and well stability and existing faults are the major risk factors in the short term. However, reservoir permeability and porosity take turns in dominating the leakage risk over the long term (e.g., 30 years). The eastern part of the study area was classified as a leakage-prone area due to the existence of high-permeability strata. Because the CO_2 leakage risk and critical risk factors dynamically change with time, a monitoring program needs to be dynamically deployed in conjunction with different time periods.

This study innovatively combined numerical simulation results with quantitative evaluation scoring indices, realizing directional feature application of the evaluation method. It also investigated dynamic relationships between the distribution of the CO_2 plume and the amount of leakage over a longer time scale. However, there is still a need for more accurate and specific descriptions of the leakage paths, such as the wellbore and faults, etc. The established method of cumulative multiplication of comprehensive indexes cannot accurately reflect the weight influence of each index on the overall leakage risk. In the future, more attention should also be paid to the deformation mechanism of fractures during the sealing process in the evaluation system.

Author Contributions: Writing—original draft, J.X.; Methodology, Formal analysis, Writing—review and editing, C.W.; Data curation, Visualization, D.L. (Dong Lin); Validation, X.W.; Writing—review and editing, Z.Z.; Writing—review and editing, Funding acquisition, Supervision, D.L. (Danqing Liu). All authors have read and agreed to the published version of the manuscript.

Funding: This research received no external funding.

Data Availability Statement: The data presented in this study are available on request from the corresponding author due to privacy.

Conflicts of Interest: Authors Jingwen Xiao, Dong Lin and Xiao Wu were employed by the PetroChina Southwest Oil & Gas Field Company. The remaining authors declare that the research was conducted in the absence of any commercial or financial relationships that could be construed as a potential conflict of interest.

Abbreviations

The following abbreviations are used in this manuscript:

CCUS	CO_2 capture, geological storage and utilization
EGR	Enhanced Gas Recovery
CASSIF	Containment Assurance Scenario and Significance Identification Framework
FEP	Features, Events, and Processes
SWIFT	Structured What-If Technique
VEF	Vulnerability Evaluation Framework
P&R	Performance and Risk
CFA	Certification Framework Approach
CO_2-PENS	CO_2-Predicting Engineered Natural Systems
RISCS	Risk Interference Subsurface CO_2 Storage

PRA	Probabilistic Risk Assessment Method
FTA	Fault Tree Analysis

References

1. Huang, W.; Xiao, Y.; Li, X.; Ning, Z. Leakage diffusion and safety assessment of CO_2 pipeline transportation flange based on CFD simulation. *Greenh. Gases* **2024**, *14*, 728–742. [CrossRef]
2. Nath, F.; Mahmood, M.N.; Yousuf, N. Recent advances in CCUS: A critical review on technologies, regulatory aspects and economics. *Geoenergy Sci. Eng.* **2024**, *238*, 212726. [CrossRef]
3. Chen, S.; Liu, J.; Zhang, Q.; Teng, F.; McLellan, B.C. A critical review on deployment planning and risk analysis of carbon capture, utilization, and storage (CCUS) toward carbon neutrality. *Renew. Sustain. Energy Rev.* **2022**, *167*, 112537. [CrossRef]
4. Ahmed, S.; Khan, M.K.; Kim, J. Revolutionary advancements in carbon dioxide valorization via metal-organic framework-based strategies. *Carbon. Capture Sci. Technol.* **2025**, *15*, 100405. [CrossRef]
5. Su, X.; Liu, S.; Zhang, L.; Liu, G.; Wang, Y.; Gan, M.; Li, X. Wellbore leakage risk management in CO_2 geological utilization and storage: A review. *Energy Rev.* **2023**, *2*, 100049. [CrossRef]
6. Lv, H.; Wang, L.; Liu, G.; Lou, Z.; Pang, L.; Cai, B.; Li, Q. Risk assessment on the CCUS project using risk breakdown structure methodology: A case study on Jilin oilfield CO_2-EOR Hei-79 block. *Greenh. Gases* **2021**, *11*, 750–763. [CrossRef]
7. Wang, F.; Cao, J.; Zhang, Y.; Aviso, K.B.; Tan, R.R.; Li, Z.; Jia, X. Safety risk assessment of the large-scale carbon capture, utilization, and storage demonstration project in Dongying, China. *J. Clean. Prod.* **2023**, *414*, 137699. [CrossRef]
8. Rui, Z.; Zeng, L.; Dindoruk, B. Challenges in the Large-Scale Deployment of CCUS. *Engineering* **2025**, *44*, 17–20. [CrossRef]
9. Wang, Z.; Li, H.; Liu, S.; Xu, J.; Liu, J.; Wang, X. Risk evaluation of CO_2 leakage through fracture zone in geological storage reservoir. *Fuel* **2023**, *342*, 127896. [CrossRef]
10. Yavuz, F.; van Tilburg, T.; David, P.; Spruijt, M.; Wildenborg, T. Second Generation CO_2 FEP Analysis: CASSIF—Carbon Storage Scenario Identification Framework. *Energy Procedia* **2009**, *1*, 2479–2485. [CrossRef]
11. Condor, J.; Unatrakarn, D.; Wilson, M.; Asghari, K. A comparative analysis of risk assessment methodologies for the geologic storage of carbon dioxide. *Energy Procedia* **2011**, *4*, 4036–4043. [CrossRef]
12. EPA, U. *Vulnerability Evaluation Framework for Geologic Sequestration of Carbon Dioxide*; Epa 430-R-08-009; U.S. Environmental Protection Agency: Washington, DC, USA, 2008.
13. Barros, E.G.D.; Leeuwenburgh, O.; Szklarz, S.P. Quantitative assessment of monitoring strategies for conformance verification of CO_2 storage projects. *Int. J. Greenh. Gas. Con.* **2021**, *110*, 103403. [CrossRef]
14. Sneddon, J.; Busby, A.; Hurst, S. Practical Qualitative and Quantitative Models in Carbon Capture Utilization and Storage Risk Assessment. In Proceedings of the Offshore Technology Conference, Houston, TX, USA, 6–9 May 2024; p. D031S042R002.
15. Le Guen, Y.; Le Gouevec, J.; Chammas, R.; Gerard, B.; Poupard, O.; Van Der Beken, A.; Jammes, L. CO_2 Storage: Managing the Risk Associated With Well Leakage Over Long Time Scales. *SPE Proj. Facil. Constr.* **2009**, *4*, 87–96. [CrossRef]
16. Oldenburg, C.M. Screening and ranking framework for geologic CO_2 storage site selection on the basis of health, safety, and environmental risk. *Environ. Geol.* **2008**, *54*, 1687–1694. [CrossRef]
17. Stauffer, P.H.; Pawar, R.J.; Surdam, R.C.; Jiao, Z.; Deng, H.; Lettelier, B.C.; Viswanathan, H.S.; Sanzo, D.L.; Keating, G.N. Application of the CO_2-PENS risk analysis tool to the Rock Springs Uplift, Wyoming. *Energy Procedia* **2011**, *4*, 4084–4091. [CrossRef]
18. Bielicki, J.M.; Pollak, M.F.; Wilson, E.J.; Fitts, J.P.; Peters, C.A. A Methodology for Monetizing Basin-Scale Leakage Risk and Stakeholder Impacts. *Energy Procedia* **2013**, *37*, 4665–4672. [CrossRef]
19. Junji Yamaguchi, A.; Sato, T.; Tobase, T.; Wei, X.; Huang, L.; Zhang, J.; Bian, J.; Liu, T. Multiscale numerical simulation of CO_2 hydrate storage using machine learning. *Fuel* **2023**, *334*, 126678. [CrossRef]
20. Sori, A.; Moghaddas, J.; Abedpour, H. Comprehensive review of experimental studies, numerical modeling, leakage risk assessment, monitoring, and control in geological storage of carbon dioxide: Implications for effective CO_2 deployment strategies. *Greenh. Gases* **2024**, *14*, 887–913. [CrossRef]
21. Lu, S.; Jiang, P.; Cong, L.; Lü, X. Research Progress on CO_2 Geological Storage Reservoir and Caprock Mechanics: Methods and Status. *Greenh. Gases* **2025**, *15*, 264–276. [CrossRef]
22. Gao, S.; Jia, S.; Zhu, Y.; Zhao, L.; Cao, Y.; Qi, X.; Guan, F. Study on Quantitative Assessment of CO_2 Leakage Risk Along the Wellbore Under the Geological Storage of the Salt Water Layer. *Energies* **2024**, *17*, 5302. [CrossRef]
23. Sorgi, C.; De Gennaro, V.; Mandiuc, A. A New Methodology for Quantitative Risk Assessment of CO_2 Leakage in CCS Projects. *Spe. J.* **2024**, *29*, 7214–7233. [CrossRef]
24. Dou, N.; Hu, Z.; Wang, Z.; He, H.; Yang, S. Risk Assessment of Annular Pressure Caused by Tubing Leakage in Offshore Gas Wells with High CO_2. *Geofluids* **2023**, *2023*, 9072217. [CrossRef]
25. Hajiyev, E.; Watson, M.; Emadi, H.; Eissa, B.; Hussain, A.; Baig, A.R.; Shahin, A. A Comparative Study of Major Risk Assessment (RA) Frameworks in Geologic Carbon Storage (GCS). *Fuels* **2025**, *6*, 42. [CrossRef]

26. Zhang, L.; Cao, C.; Wen, S.; Zhao, Y.; Peng, X.; Wu, J. Thoughts on the development of CO_2-EGR under the background of carbon peak and carbon neutrality. *Nat. Gas. Ind. B* **2023**, *10*, 383–392. [CrossRef]
27. Li, Q.; Liu, G. Risk Assessment of the Geological Storage of CO_2: A Review. In *Geologic Carbon Sequestration: Understanding Reservoir Behavior*; Vishal, V., Singh, T.N., Eds.; Springer International Publishing: Cham, Switzerland, 2016; pp. 249–284.
28. Zhang, Y.; Oldenburg, C.M.; Benson, S.M. Vadose Zone Remediation of Carbon Dioxide Leakage from Geologic Carbon Dioxide Sequestration Sites. *Vadose Zone J.* **2004**, *3*, 858–866. [CrossRef]
29. Gholami, R.; Raza, A.; Iglauer, S. Leakage risk assessment of a CO_2 storage site: A review. *Earth-Sci. Rev.* **2021**, *223*, 103849. [CrossRef]
30. Jing, M.; Li, Q.; Liu, G.; Xue, Q. Quantitative assessment of CO_2 leakage risk in geologic carbon storage management. *Greenh. Gases* **2024**, *14*, 1068–1091. [CrossRef]
31. Jia, R.; Fu, X.; Jin, Y.; Wu, T.; Wang, S.; Cheng, H. Mechanical properties of mudstone caprock and influencing factors: Implications for evaluation of caprock integrity. *Front. Earth Sci.* **2023**, *11*, 1229851. [CrossRef]
32. Xiao, T.; Xu, H.; Moodie, N.; Esser, R.; Jia, W.; Zheng, L.; Rutqvist, J.; McPherson, B. Chemical-Mechanical Impacts of CO_2 Intrusion Into Heterogeneous Caprock. *Water Resour. Res.* **2020**, *56*, e2020WR027193. [CrossRef]
33. Watson, T.L.; Bachu, S. Identification of Wells with High CO_2-Leakage Potential in Mature Oil Fields Developed for CO_2-Enhanced Oil Recovery. In Proceedings of the SPE Symposium on Improved Oil Recovery, Tulsa, OK, USA, 20–23 April 2008; p. SPE-112924-MS.
34. Lackey, G.; Vasylkivska, V.S.; Huerta, N.J.; King, S.; Dilmore, R.M. Managing well leakage risks at a geologic carbon storage site with many wells. *Int. J. Greenh. Gas. Con.* **2019**, *88*, 182–194. [CrossRef]
35. Lu, J.; Lv, Y.; Fang, X.; Zuo, J.; Wang, S.; Li, H.; Yuan, C.; Liu, W. Fault Leakage Behaviors and Co_2 Migration in Different Types of Geological Carbon Storage. *Chem. Technol. Fuels Oils* **2024**, *60*, 471–480. [CrossRef]
36. Fu, X.; Jia, R.; Wang, H.; Wu, T.; Meng, L.; Sun, Y. Quantitative evaluation of fault-caprock sealing capacity: A case from Dabei-Kelasu structural belt in Kuqa Depression, Tarim Basin, NW China. *Pet. Explor. Dev.* **2015**, *42*, 329–338. [CrossRef]
37. Xu, J.; Zhou, Y.; Pan, R.; Huang, T.; Liu, B.; Gao, Q.; Liu, Y. The Integrity Evaluation of Geological Body for CCUS Technology Based on Microseismic Monitoring Technology. In Proceedings of the ADIPEC, Abu Dhabi, United Arab Emirates, 4–7 November 2024; p. D021S061R004.
38. Eaton, B.A. Fracture Gradient Prediction and Its Application in Oilfield Operations. *J. Pet. Technol.* **1969**, *21*, 1353–1360. [CrossRef]
39. Liu, D.; Agarwal, R.; Liu, F.; Yang, S.; Li, Y. Modeling and assessment of CO_2 geological storage in the Eastern Deccan Basalt of India. *Environ. Sci. Pollut. Res.* **2022**, *29*, 85465–85481. [CrossRef]

MDPI AG
Grosspeteranlage 5
4052 Basel
Switzerland
Tel.: +41 61 683 77 34

Energies Editorial Office
E-mail: energies@mdpi.com
www.mdpi.com/journal/energies

www.ingramcontent.com/pod-product-compliance
Lightning Source LLC
Chambersburg PA
CBHW061617210926
51850CB00009B/2355
* 9 7 8 3 7 2 5 8 5 1 2 9 4 *